Foundations in Signal Processing, Communications and Networking

Series Editors: W. Utschick, H. Boche, R. Mathar

For other titles published in this series, go to
www.springer.com/series/7603

Johannes Brehmer

Utility Maximization in Nonconvex Wireless Systems

Springer

Johannes Brehmer
Technische Universität München
Associate Institute for Signal Processing
80290 Munich
Germany

ISSN 1863-8538 e-ISSN 1863-8546
ISBN 978-3-642-17437-7 e-ISBN 978-3-642-17438-4
DOI 10.1007/978-3-642-17438-4
Springer Heidelberg Dordrecht London New York

Library of Congress Control Number: 2012934555

Printed on acid-free paper

Springer is part of Springer Science+Business Media (www.springer.com)

Contents

Chapter 1
Introduction

Once upon a time, cellular wireless networks provided two basic services: voice telephony and low-rate text messaging. Users in the network were separated by orthogonal multiple access schemes, and cells by generous frequency reuse patterns [1]. Since then, the proliferation of wireless services, fierce competition, and the emergence of new service classes such as wireless data and multimedia have resulted in an ever increasing pressure on network operators to use resources in a more efficient manner. In the context of wireless networks, two of the most common resources are power and spectrum—and, due to regulations, these resources are typically scarce. Hence, in contrast to wired networks, overprovisioning is not feasible in wireless networks. Instead, every network operator faces a resource allocation problem—to maximize revenue under the constraint of limited resources. In a voice-only system, maximizing revenue is equivalent to maximizing the number of users in the system. Intuitively, the problem becomes more intricate with the introduction of different service classes.

A key technology for increasing overall system performance are *advanced physical layer techniques*. These techniques can be categorized into three groups: multi-antenna techniques, multi-carrier techniques, and multi-user techniques. Multi-antenna techniques were first used in single-link (point-to-point) transmission, by equipping either receiver, or transmitter, or both with multiple antenna elements. The popularity of multi-carrier techniques can be attributed to their ability to exploit frequency and multi-user diversity in a highly efficient manner. Multi-user technologies do away with the paradigm of orthogonal multiple access. By allowing for interference, they have the potential to significantly increase system performance. This potential, however, can only be unleashed by proper management of interference. In its most basic form, interference management corresponds to a power control problem [2]. A straightforward approach is to treat interference as noise. More elaborate multi-user coding techniques allow for decoding of received signals as if part of the interference was not present. Further gains are possible by a synthesis of multi-antenna and multi-user techniques. By using multiple antennas, interference management can operate in the spatial

J. Brehmer, *Utility Maximization in Nonconvex Wireless Systems*, Foundations in Signal Processing, Communications and Networking 5, DOI 10.1007/978-3-642-17438-4_1,

dimension, which basically enables the network operator to serve multiple users at the same time in the same frequency band, in a power-efficient way. By allowing more than one user to access the wireless medium at the same time and at the same frequency, the traditional separation between *medium access control* (MAC) layer and physical layer is dissolved. Based on this observation, a rather broad definition of the term physical layer is adopted, which does not distinguish between physical and MAC layer.

While advanced physical layer techniques have the potential to increase overall system performance, they also further complicate the task to operate the system in an optimal manner. In other words, advanced techniques add new degrees of freedom to the resource allocation problem, but using these degrees of freedom in a beneficial manner is a highly nontrivial task. Simply speaking, advanced physical layer techniques further complicate the resource allocation problem. Naturally, most of the aforementioned techniques add a number of tunable parameters to the resource allocation problem. It is important to realize, however, that the impact of advanced physical layer techniques goes far beyond adding a few more parameters. As an example, multi-user techniques generally imply a structure of the resource allocation problem that is fundamentally different from that of a simple system based on orthogonal multiple access, due to the fact that interference generally leads to nonconvex optimization problems [3].

Furthermore, advanced physical layer techniques are not the only source of complication in the resource allocation problem. Originally, there was a single application in cellular wireless systems, voice telephony. In recent years, however, the technological premises for wireless data traffic have been created, and wireless data has become a generator of significant revenue [4]. In its simplest form, voice traffic can be modeled as inelastic: the voice codec needs a minimum rate, below this rate transmission fails, above this rate no further increase in voice quality results. In contrast, data traffic is elastic: under the assumption that a document (e.g., web page, email, video clip) of a given size is to be transmitted, there is no minimum rate requirement, but smaller rate results in higher delay, hence a larger rate is desirable. Given this flexibility of data services, criteria for rate allocation are needed. Rate allocation among users is a well-known problem in rate control for the internet [5]. The most common criteria for rate allocation are system throughput (i.e., sum rate), max–min fairness, and proportional fairness [5]. The optimum rate allocation can be obtained by solving a utility maximization problem, under the assumption that the objectives throughput, max–min fairness, and proportional fairness can be represented in terms of concave and increasing utility functions. All three objectives have in common that they do not take into account what the transmitted data is actually used for. In other words, they are blind with respect to the end-user applications that are fed by the transmitted data. As an example, one user may surf the web, while another receives an email. An obvious next step is to generalize the notion of a utility function and let the utility function be any function that captures the properties of the end-user application. By using such generalized utility functions, the resource allocation problem becomes *application-oriented* [6, 7]. While it is still reasonable to assume that performance of the end-user application increases with the allocated rate, it is no longer obvious that a generalized utility function should be concave [6]. As an example, a video codec may need a minimum rate to operate. Or, if the video codec is a layered codec [8], the finite number of layers leads to a staircase-like utility function [7]. As a result, it may be possible to obtain a more precise model of the application by using a nonconcave utility function. Moreover, each user in the system may use a different application. Accordingly, the overall system utility may be a combination of utilities of different types and shapes, some concave, some not. A generalized utility function already represents an abstraction of an application: It describes how the application reacts to a given rate. A more detailed model of an application may include a set of tunable application parameters, such as source coding rate or number of encoding layers.

Evidently, there are two trends in wireless networks whose convergence poses significant challenges to the network operator: First, in order to increase physical layer performance, advanced physical layer techniques are necessary. Second, applications become adaptable and more diverse. As a result, in order to translate the performance potential of advanced physical layer techniques into increased overall system performance, precise models of the applications are required. Due to the coupling between layers, a joint optimization across layers is required (*cross-layer optimization*). But achieving optimal system performance is far from trivial, due to the intricate structure of the resulting optimization problem. In particular, *nonconvexities* occur at both ends of the layered architecture: At the physical layer due to interference, and at the application layer due to precise models of the actual applications.

Cross-layer optimization has been a popular field of research in recent years, resulting in a wealth of results and a huge body of literature on the topic, see, e.g., [9–16] and references therein. By definition, cross-layer optimization is a multi-disciplinary topic, spanning disciplines such as networking, physical layer signal processing, and multimedia signal processing. The multi-disciplinary character creates a need for unifying frameworks, which allow for a modularization of the overall problem into a set of coupled subproblems, such that each subproblem can be addressed by experts from within each discipline. Moreover, such a unifying framework provides a common ground, which helps to divide fundamental properties from artifacts. The wealth of piecemeal approaches in the cross-layer optimization literature underlines the importance of such frameworks. One such framework is *network utility maximization* (NUM) [14]. The fundamental achievement of this framework is that it recognizes cross-layer optimization as a mathematical optimization problem with a decomposable structure. Moreover, the optimization problem is not solved by heuristics or exhaustive search, but by state-of-the-art numerical methods. Based on this foundation on mathematical methods, it is possible to model, analyze, and solve cross-layer problems with respect to their mathematical properties.

A NUM problem was first formulated in the context of rate control for wired networks [5], under the assumption that link capacities are fixed. In wireless networks, link capacities are "flexible" [17], as they depend on the underlying resource allocation at the physical layer [17, 18]. Including the optimization of the physical layer resource allocation into the NUM problem results in a generalized NUM problem [14]. In case the physical layer parameters are link transmission powers, NUM is equivalent to utility-based power control. Most works on NUM assume that the utility maximization problem can be cast as a convex optimization problem. Only a few works have considered nonconvex problem setups [19]. In case of NUM with fixed link capacities the convexity assumption implies concave utility functions. For generalized NUM, the assumption of problem convexity additionally implies a convex model of the physical layer.[1] In conjunction with advanced physical layer techniques, such a convexity assumption is difficult to fulfill. In the special case that the physical layer only performs power control, it is possible to obtain a convex model based on a suitable reformulation, either by using a high-SIR approximation [17, 18] or by restricting the class of admissible utility models [18]. For general application and physical layer models, however, such a convex reformulation is generally not possible. An alternative to convex reformulation is to avoid convexity from the beginning, by choosing a convex model for the problem at hand ("modeling for optimization", [14]). Unfortunately, when using advanced physical layer techniques, it is often not possible to obtain a convex model of the physical layer that is sufficiently precise. Similarly, restricting utility models to concave functions conflicts with the paradigm of using application-oriented utilities [6, 7]. As a consequence, the trend to advanced physical layer techniques and application-oriented utilities makes it impossible to avoid nonconvex problems. Instead, it is essential to confront nonconvex problems and to develop strategies to solve such problems within a well-defined mathematical framework.

The focus on nonconvex problems requires modesty with respect to the size of networks and the properties of algorithms for solving the problems. The convexity assumption in classic NUM is no coincidence: Convexity can be considered a prerequisite for the development of distributed algorithms for large networks [3]. The inherent complexity of global methods for nonconvex optimization [20] limits their applicability to simple network topologies. Moreover, all methods developed in this monograph are centralized algorithms, i.e., they cannot be implemented in a distributed manner. Finally, the complexity of the global methods prevents their usage in on-line applications. In other words, even in a wireless network with a centralized controller, it is generally not possible to solve a nonconvex resource allocation problem to global optimality in real-time.

Despite their limitations with respect to practical implementation, global methods provide valuable results, from both a theoretical and a practical perspective. Using global methods, it is possible to explore the ultimate performance bounds for a given system configuration. Furthermore, the global optimum represents a benchmark for any practically implementable method. While it is interesting to see that heuristic a performs better than heuristic b, a true assessment of the quality of any proposed method is only possible if the global optimum is known.

---

[1] A precise definition of a physical layer model is provided in Chap. 2.

There exist multiple ways to obtain suboptimal, but practically implementable solutions for a nonconvex problem. One possible way is to develop a method based on "engineering intuition". An alternative path is to relax the optimality criterion, and accept parameter setups that fulfill a local optimality criterion as solutions to a nonconvex problem. Local methods have the advantage that the solution still fulfills some optimality criterion, while it is usually difficult to make statements about the theoretical properties of a purely "engineered" solution. Still, the two ways need not be antipodes. Ideally, engineering intuition and mathematical rigor go hand in hand in the development of powerful methods.

This monograph can be divided in two parts, theory and application. Chapters 2 and 3 provide the theoretical framework for modeling and solving utility maximization problems in a wireless system, including both advanced physical layer techniques and application-oriented utility functions. In Chap. 2, a model for parameter optimization is developed. Intentionally, the model is general enough to encompass a wide array of system configurations and performance objectives. In particular, no assumptions concerning problem convexity are made. Chapter 3 discusses various methods for solving a given utility maximization problem. The applicability of each solution method depends on mathematical properties of the utility maximization problem. In other words, each solution method is applicable to a class of problems. Chapters 4–6 demonstrate the application of the proposed framework. Chapter 4 provides detailed physical layer models for multi-antenna broadcast channels, while Chap. 5 discusses some common utility models. In Chap. 6 physical layer models and utility models are combined to yield instances of the utility maximization problem defined in Chap. 2. These instances are then solved using the methods from Chap. 3.

Chapter 2
General Problem Setup

The functionality provided by a wireless communication system can usually be partitioned into submodules, the so-called layers [21]. With each layer, a set of variable parameters can be associated. For example, common physical layer parameters are power allocation, allocation of time slots, beamforming vectors, choice of modulation and coding schemes, etc. Possible parameters at upper layers are scheduling decision, choice of a route, and, at the application layer, source coding rate. Parameters are variable in the sense that for each parameter there exists a set of possible values to choose from. Moreover, each parameters impacts overall system performance in a certain way. Generally speaking, then, a parameter optimization corresponds to optimizing system performance with respect to all variable parameters over the set of all possible parameter setups.

This chapter starts with the definition of a general model of a parameter optimization problem in a layered wireless communication system. In a second step, the properties of the layers above the physical layer are condensed in a so-called utility model. The resulting utility maximization problem represents the basic model that is used throughout this monograph. Due to its generality, the model covers a wide array of system configurations and performance metrics.

In later chapters, the abstractions used in the general model will be refined, resulting in specific instances of the utility optimization problem. Starting with a general model helps to distinguish between generic properties of the class of utility maximization problems and the algorithms used to solve it, and properties that are specific to each problem instance.

2.1 Parameter Optimization Problems

The OSI reference model [21] defines a model for partitioning the functionality of a communication system into a hierarchy of subfunctions. In the definition of the parameter optimization problem, a similar model will be used. The focus, however,

J. Brehmer, *Utility Maximization in Nonconvex Wireless Systems*, Foundations in Signal Processing, Communications and Networking 5, DOI 10.1007/978-3-642-17438-4_2,

Fig. 2.1 Transmission of scalable video data from base station to two receivers

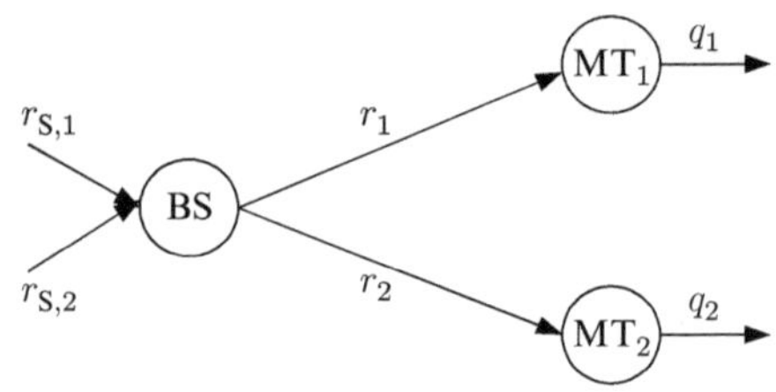

is not on the functionality of a system, but on how the different parameters in the system impact the overall performance of the system. Moreover, based on the observation that in a wireless system, nonconvexities primarily occur at the physical layer and at the application layer, the development is based on a two-layer model.

Example 2.1.1. As an introductory example, consider the communication scenario depicted in Fig. 2.1:

Video data is sent from a wireless base station to two mobile receivers, where it is assumed that the mobile terminals receive different video streams. The transmit power of the base station is limited to a maximal value $P > 0$. Assume that the only parameters at the physical layer are the allocated transmit powers to the users. Let p_k denote the power allocated to user k. The set of feasible transmit power allocations $\boldsymbol{p} = (p_1, p_2)$ is then given by

$$\mathcal{P} = \left\{\boldsymbol{p} \in \mathbb{R}_+^2 : \mathbf{1}^{\mathrm{T}}\boldsymbol{p} \leq P\right\}. \tag{2.1}$$

It is assumed that each transmit power allocation results in a transmission rate $r_k(\boldsymbol{p})$ to user k that can be reliably provided. The details of the rate function $\boldsymbol{r} : \mathbb{R}_+^2 \to \mathbb{R}^2$ that maps a transmit power allocation $\boldsymbol{p}$ into a rate vector $\boldsymbol{r}(\boldsymbol{p})$ are not specified further at this point, it is simply assumed that $\boldsymbol{r}$ is continuous and that

$$\boldsymbol{r}(\mathbf{0}) = \mathbf{0}. \tag{2.2}$$

The video streams are encoded with a layered scalable video codec such as H.264 SVC [8]. Accordingly, the base station can adapt the source rate of each video stream to the capabilities of the physical layer by transmitting a subset of layers. Let L_k denote the maximum number of layers of stream k. Furthermore, let $\ell_k \in \{0, \ldots, L_k\}$ denote the actual number of video layers transmitted to user k, and $r_{S,k}(\ell_k)$ the resulting source rate. If no layer is transmitted, the corresponding source rate is zero:

$$r_{S,k}(0) = 0, k = 1, 2. \tag{2.3}$$

It is assumed that a video stream can be successfully transmitted if its source rate is smaller than or equal to the rate provided by the physical layer. This results in the following constraint on the rates:

$$r_{S,k}(\ell_k) \le r_k(\boldsymbol{p}), k = 1, 2. \tag{2.4}$$

Each choice of the number of transmitted layers ℓ_k results in a certain quality of the decoded video stream at receiver k, described by a function $q_k : \{1, \dots, L_k\} \to \mathbb{R}, \ell_k \mapsto q_k(\ell_k)$.

In this exemplary scenario, the parameters to adjust are the transmit power allocation $\boldsymbol{p}$ and the number of video layers $\boldsymbol{\ell} = (\ell_1, \ell_2)$. The objective of the system is to maximize the sum quality of service $q(\boldsymbol{\ell}) = q_1(\ell_1) + q_2(\ell_2)$ over the set of feasible parameter values:

$$\max_{\boldsymbol{\ell}, \boldsymbol{p}} q(\boldsymbol{\ell}) \quad \text{s.t.} \quad (\boldsymbol{\ell}, \boldsymbol{p}) \in \mathcal{L} \times \mathcal{P}, \boldsymbol{r}_S(\boldsymbol{\ell}) \le \boldsymbol{r}(\boldsymbol{p}), \tag{2.5}$$

where

$$\mathcal{L} = \{0, \dots, L_1\} \times \{0, \dots, L_2\} \subset \mathbb{R}^2.$$

The maximum exists: According to (2.2) and (2.3), the set of feasible parameters

$$\mathcal{F} = \{(\boldsymbol{\ell}, \boldsymbol{p}) \in \mathcal{L} \times \mathcal{P} : \boldsymbol{r}_S(\boldsymbol{\ell}) \le \boldsymbol{r}(\boldsymbol{p})\} \tag{2.6}$$

is nonempty. Moreover, $\mathcal{L}$ and $\mathcal{P}$ are compact, and both $\boldsymbol{r}_S$ and $\boldsymbol{r}$ are continuous, hence $\mathcal{F}$ is compact.[1] Finally, q is continuous. As a result, there exists $(\boldsymbol{\ell}^*, \boldsymbol{p}^*) \in \mathcal{F}$ such that

$$q(\boldsymbol{\ell}^*) = \sup_{(\boldsymbol{\ell}, \boldsymbol{p}) \in \mathcal{F}} q(\boldsymbol{\ell}) < \infty. \tag{2.7}$$

After defining a function $\boldsymbol{g}$, with

$$\boldsymbol{g}(\boldsymbol{y}, \boldsymbol{z}) = \boldsymbol{z} - \boldsymbol{y}, \tag{2.8}$$

the constraint (2.4) can be written as

$$\boldsymbol{g}(\boldsymbol{r}_S(\boldsymbol{\ell}), \boldsymbol{r}(\boldsymbol{p})) \ge \boldsymbol{0}. \tag{2.9}$$

Note that the function $\boldsymbol{g}(\boldsymbol{y}, \cdot)$ is increasing. Such a monotonicity in the performance metrics of the physical layer is highly desirable: Basically, a better performance of the physical layer should never reduce the set of feasible parameter setups at the upper layers—otherwise it could happen that better performance at the physical

---

[1] $\mathcal{L}$ is a finite subset of $\mathbb{R}^2$. Accordingly, any function $f : \mathcal{L} \to \mathbb{R}$ is continuous. This result follows by considering the subspace topology in $\mathcal{L}$—intersecting the open sets in $\mathbb{R}^2$ with $\mathcal{L}$ yields the open sets in $\mathcal{L}$. Obviously, any subset of $\mathcal{L}$ is an open set (the discrete topology). As a result, the preimage of any open set $I \subseteq \mathbb{R}$ under f is open, showing that f is continuous, cf. Proposition A.2.1.

layer leads to reduced overall performance, or, in the worst case, infeasibility, a behaviour that is clearly undesirable.

Example 2.1.2. A slight change in the Example 2.1.1 results in a different optimization problem: The sum quality of the decoded video can also be modeled as a function of the source rate, $q : \mathbb{R}^2 \to \mathbb{R}, \boldsymbol{r}_S \mapsto q(\boldsymbol{r}_S)$. The following assumptions concerning q are made: q is continuous and increasing, i.e., a higher source rate does not result in a reduced quality of the decoded stream. Due to the monotonicity of q, given a rate $\boldsymbol{r}(\boldsymbol{p})$ from the physical layer, the optimal strategy to choose the source rate is obviously to set

$$\boldsymbol{r}_S = \boldsymbol{r}(\boldsymbol{p}). \tag{2.10}$$

The resulting optimization problem now reads as:

$$\max_{\boldsymbol{p}} q(\boldsymbol{r}(\boldsymbol{p})) \quad \text{s.t.} \quad \boldsymbol{p} \in \mathcal{P}. \tag{2.11}$$

Similar to Problem (2.5), the existence of the maximum is now ensured by the compactness of $\mathcal{P}$ and the continuity of $\boldsymbol{r}$: With q and $\boldsymbol{r}$ continuous, the composite function $q \circ \boldsymbol{r}$ is continuous. As a result, from the compactness of $\mathcal{P}$ it follows that $q(\boldsymbol{r}(\mathcal{P}))$ is compact, thus the maximum exists. In addition, there exists a power allocation $\boldsymbol{p}^* \in \mathcal{P}$ that achieves the maximum.

Note that the compactness of $q(\boldsymbol{r}(\mathcal{P}))$ can also be ensured by making additional structural assumptions on q—in this work, however, all models are restricted to compact parameter sets and continuous performance maps.

Example 2.1.3. A variation of Example 2.1.2 is obtained by requiring a minimum (source) rate for each user:

$$\max_{\boldsymbol{p}} q(\boldsymbol{r}(\boldsymbol{p})) \quad \text{s.t.} \quad \boldsymbol{p} \in \mathcal{P}, \boldsymbol{g}(\boldsymbol{r}(\boldsymbol{p})) \geq \boldsymbol{0}. \tag{2.12}$$

with $\boldsymbol{g} : \mathbb{R}^2 \to \mathbb{R}^2$,

$$\boldsymbol{g}(\boldsymbol{y}) = \boldsymbol{y} - \boldsymbol{r}_{\min}. \tag{2.13}$$

Adding a minimum rate constraint leads to a feasibility problem: If $\boldsymbol{r}_{\min} > \boldsymbol{0}$, it is not clear whether Problem (2.12) is feasible—i.e., the structural assumptions made on $\mathcal{P}$ and $\boldsymbol{r}$ are not sufficient to establish the existence of the maximum.

Note that besides compactness and continuity, a major structural property present in Problem (2.11) is again monotonicity: both q and $\boldsymbol{g}$ are increasing functions.

Examples 2.1.1–2.1.3 each result in an optimization problem. From the perspective of the structural properties offered by each of these problems, they are, quite naturally, very similar. In other words, they represent particular instances of a more general optimization problem. In the following, a generic *parameter optimization problem* is defined. This generic problem contains Examples 2.1.1–2.1.3 as special cases.

Definition 2.1.1 (Parameter Optimization Problem). Let $\mathcal{D}_1$, $\mathcal{D}_2$, and $\mathcal{D}_3$ be metric spaces. Let $\mathcal{X}_1 \subseteq \mathcal{D}_1$ and $\mathcal{X}_2 \subseteq \mathcal{D}_2$ be compact, and let $\boldsymbol{r} : \mathcal{D}_1 \to \mathbb{R}^K$, $q : \mathcal{D}_2 \times \mathcal{D}_3 \to \mathbb{R}$, and $\boldsymbol{g} : \mathcal{D}_2 \times \mathcal{D}_3 \to \mathbb{R}^M$ denote three continuous functions, for some K and M, with $\boldsymbol{r}(\mathcal{X}_1) \subseteq \mathcal{D}_3$. Then

$$\sup_{\boldsymbol{x}_1,\boldsymbol{x}_2} q(\boldsymbol{x}_2, \boldsymbol{r}(\boldsymbol{x}_1)) \quad \text{s.t.} \quad \boldsymbol{x}_1 \in \mathcal{X}_1, \boldsymbol{x}_2 \in \mathcal{X}_2, \boldsymbol{g}(\boldsymbol{x}_2, \boldsymbol{r}(\boldsymbol{x}_1)) \geq \boldsymbol{0}. \tag{2.14}$$

is a *parameter optimization problem* if the following two additional conditions hold:

(1) The set of feasible parameters is nonempty, i.e.,

$$\mathcal{F} = \{(\boldsymbol{x}_1, \boldsymbol{x}_2) \in \mathcal{X}_1 \times \mathcal{X}_2 : \boldsymbol{g}(\boldsymbol{x}_2, \boldsymbol{r}(\boldsymbol{x}_1)) \geq \boldsymbol{0}\} \neq \emptyset. \tag{2.15}$$

(2) The functions $q(\boldsymbol{y}, \cdot)$ and $\boldsymbol{g}(\boldsymbol{y}, \cdot)$ are increasing.

In the general model, the variable $\boldsymbol{x}_1$ contains all adaptable parameters of the physical layer, and $\mathcal{X}_1$ represents the set of all possible physical layer parameter setups. Likewise, $\boldsymbol{x}_2$ contains the parameters of all upper layers, and $\mathcal{X}_2$ is the set of all possible parameter setups at the upper layers. The function $\boldsymbol{r}$ maps a physical layer parameter setup into a vector of K physical layer performance metrics. The function $\boldsymbol{g}$ maps a parameter setup of the upper layers and a vector of physical layer performance metrics into a vector of M upper layer performance metrics. Accordingly, the function $\boldsymbol{g}$ can be used to impose M minimum performance requirements. Finally, the function q maps a parameter setup of the upper layers and a vector of physical layer performance metrics into a single overall performance metric. Accordingly, the parameter optimization problem corresponds to maximizing the overall performance metric q with respect to the parameters of all layers, under some minimum performance requirements.

Closely related is the problem of determining an optimum parameter setup, i.e., a choice of parameters $(\boldsymbol{x}_1^*, \boldsymbol{x}_2^*) \in \mathcal{X}_1 \times \mathcal{X}_2$ such that

$$q(\boldsymbol{x}_2^*, \boldsymbol{r}(\boldsymbol{x}_1^*)) = \sup\{q(\boldsymbol{x}_2, \boldsymbol{r}(\boldsymbol{x}_1)) : (\boldsymbol{x}_1, \boldsymbol{x}_2) \in \mathcal{X}_1 \times \mathcal{X}_2, \boldsymbol{g}(\boldsymbol{x}_2, \boldsymbol{r}(\boldsymbol{x}_1)) \geq \boldsymbol{0}\}. \tag{2.16}$$

Proposition 2.1.1 shows that such a parameter setup exists.

Proposition 2.1.1. *Let* $(q, \boldsymbol{g}, \boldsymbol{r}, \mathcal{X}_2, \mathcal{X}_1)$ *be the data of a parameter optimization problem according to Definition 2.1.1. Then there exists an optimal parameter setup, i.e., a parameter setup* $(\boldsymbol{x}_1^*, \boldsymbol{x}_2^*) \in \mathcal{X}_1 \times \mathcal{X}_2$ *fulfilling* (2.16).

Proof. The function $\boldsymbol{g} \circ (\mathrm{id}, \boldsymbol{r})$ is continuous. According to Proposition A.2.1, the preimage of $\mathbb{R}_+^M$ under $\boldsymbol{g} \circ (\mathrm{id}, \boldsymbol{r})$ is closed. By Proposition A.2.3, the set of feasible parameter setups

$$\mathcal{F} = (\boldsymbol{g} \circ (\mathrm{id}, \boldsymbol{r}))^{-1}\left(\mathbb{R}_+^M\right) \cap (\mathcal{X}_2 \times \mathcal{X}_1)$$

is compact. By Definition 2.1.1, $\mathcal{F}$ is nonempty. The function $q \circ (\mathrm{id}, \boldsymbol{r})$ is continuous, which, by Theorem A.2.4, yields the result. □

Considering the proof of Proposition 2.1.1, (2.15), the continuity of $\boldsymbol{r}$, q, and $\boldsymbol{g}$, and the compactness of $\mathcal{X}_1$ and $\mathcal{X}_2$ ensure that there exists a parameter setup $(\boldsymbol{x}_1, \boldsymbol{x}_2) \in \mathcal{X}_1 \times \mathcal{X}_2$ that achieves the maximal system performance. Accordingly, these properties can be considered as basic properties which guarantee that searching for an optimal parameter setup is a well-posed problem.

In addition to the property that an optimal parameter setup exists *by definition*, a parameter optimization in the sense of Definition 2.1.1 differs from an arbitrary optimization problem in mainly two aspects: the nested structure of both the objective function $q(\cdot, \boldsymbol{r}(\cdot))$ and the constraint function $\boldsymbol{g}(\cdot, \boldsymbol{r}(\cdot))$, and the monotonicity of the functions $q(\boldsymbol{y}, \cdot)$ and $\boldsymbol{g}(\boldsymbol{y}, \cdot)$. As discussed in Chap. 3, the nested structure of the parameter optimization problem plays a key role in the development of methods for computing an optimum parameter setup. Monotonicity was already motivated in Examples 2.1.1 and 2.1.3. In a layered communication system, it is clearly a desirable behavior that a performance improvement at the physical layer does not lead to a performance degradation at higher layers.

In this work, it is assumed that all performance metrics express some kind of utility, i.e., for any performance metric, a larger value is desirable. In general, physical layer performance is measured by more than a single metric. Accordingly, each vector in $\boldsymbol{r}(\mathcal{X}_1)$ represents an achievable vector of physical layer performance metrics. Let $\boldsymbol{s}, \boldsymbol{s}' \in \boldsymbol{r}(\mathcal{X}_1)$. Following the convention that larger values are more desirable, if $\boldsymbol{s} \geq \boldsymbol{s}'$ then $\boldsymbol{s}$ corresponds to a physical layer performance that is at least as good as the physical layer performance represented by $\boldsymbol{s}'$. The set of feasible parameters of the upper layers for a given physical layer performance vector $\boldsymbol{s}$ follows from the minimum performance requirements at the upper layers:

$$\mathcal{F}_2(\boldsymbol{s}) = \{\boldsymbol{x}_2 \in \mathcal{X}_2 : \boldsymbol{g}(\boldsymbol{x}_2, \boldsymbol{s}) \geq \boldsymbol{0}\}. \tag{2.17}$$

Note that $\mathcal{F}_2(\boldsymbol{s})$ may be empty—in this case, the performance vector $\boldsymbol{s}$ provided by the physical layer is not sufficient to satisfy the requirements at the upper layers. Define a function $u : \mathcal{D}_3 \to \{-\infty\} \cup \mathbb{R}$ as follows[2]:

$$u(\boldsymbol{s}) = \sup_{\boldsymbol{x}_2 \in \mathcal{F}_2(\boldsymbol{s})} q(\boldsymbol{x}_2, \boldsymbol{s}). \tag{2.18}$$

The function u corresponds to the best overall performance that can be achieved under the constraint that the physical layer provides a performance vector $\boldsymbol{s}$. A value $u(\boldsymbol{s}) = -\infty$ indicates that $\mathcal{F}_2(\boldsymbol{s}) = \emptyset$, i.e., the upper layer requirements cannot

[2] The function u is defined using the set of extended real numbers. A definition of the set of extended real numbers is provided in Appendix A.6.

be satisfied for the given physical layer performance vector. In order to guarantee that improved performance at the physical layer does not lead to a degradation in overall system performance, it is required that the function u is increasing. The monotonicity of the functions $q(\boldsymbol{y}, \cdot)$ and $\boldsymbol{g}(\boldsymbol{y}, \cdot)$ ensures just this behaviour, as shown by Proposition 2.1.2.

Lemma 2.1.1. *The set $\mathcal{F}_2(\boldsymbol{s})$ is compact.*

Proof. The function $\boldsymbol{g}(\cdot, \boldsymbol{s})$ is continuous, hence the preimage of $\mathbb{R}_+^M$ under $\boldsymbol{g}(\cdot, \boldsymbol{s})$ is closed. By Proposition A.2.3,

$$\mathcal{F}_2(\boldsymbol{s}) = \boldsymbol{g}(\cdot, \boldsymbol{s})^{-1}(\mathbb{R}_+^M) \cap \mathcal{X}_2$$

is compact. □

Proposition 2.1.2. *The function u is increasing.*

Proof. Let $\boldsymbol{s} \in \mathcal{D}_3$ and $\boldsymbol{s} \leq \boldsymbol{s}'$. If $\mathcal{F}_2(\boldsymbol{s}) = \emptyset$, then $u(\boldsymbol{s}) = -\infty$ and the result trivially holds. Now assume that $\mathcal{F}_2(\boldsymbol{s})$ is nonempty. From the monotonicity of $\boldsymbol{g}(\boldsymbol{y}, \cdot)$, it follows that

$$\mathcal{F}_2(\boldsymbol{s}) \subseteq \mathcal{F}_2(\boldsymbol{s}'), \tag{2.19}$$

as $\boldsymbol{g}(\boldsymbol{x}_2, \boldsymbol{s}) \geq \boldsymbol{0}$ implies $\boldsymbol{g}(\boldsymbol{x}_2, \boldsymbol{s}') \geq \boldsymbol{0}$. Moreover, the monotonicity of $q(\boldsymbol{y}, \cdot)$ implies that

$$q(\boldsymbol{x}_2, \boldsymbol{s}') \geq q(\boldsymbol{x}_2, \boldsymbol{s}), \forall \boldsymbol{x}_2 \in \mathcal{X}_2,$$

thus

$$\max_{\boldsymbol{x}_2 \in \mathcal{F}_2(\boldsymbol{s})} q(\boldsymbol{x}_2, \boldsymbol{s}') \geq \max_{\boldsymbol{x}_2 \in \mathcal{F}_2(\boldsymbol{s})} q(\boldsymbol{x}_2, \boldsymbol{s}). \tag{2.20}$$

As a result,

$$u(\boldsymbol{s}') = \max_{\boldsymbol{x}_2 \in \mathcal{F}_2(\boldsymbol{s}')} q(\boldsymbol{x}_2, \boldsymbol{s}') \geq \max_{\boldsymbol{x}_2 \in \mathcal{F}_2(\boldsymbol{s})} q(\boldsymbol{x}_2, \boldsymbol{s}') \geq \max_{\boldsymbol{x}_2 \in \mathcal{F}_2(\boldsymbol{s})} q(\boldsymbol{x}_2, \boldsymbol{s}) = u(\boldsymbol{s}),$$

where the first inequality follows from (2.19), and the second inequality is just (2.20). □

To summarize, the properties of the physical layer are completely characterized by the pair $(\mathcal{X}_1, \boldsymbol{r})$. The function u describes how the upper layers react (in terms of overall system performance) to a rate vector $\boldsymbol{s} = \boldsymbol{r}(\boldsymbol{x}_1)$ provided by the physical layer. The components of a rate vector $\boldsymbol{r}(\boldsymbol{x}_1)$ represent the metrics in which the performance of the physical layer is measured. Proposition 2.1.2 shows that overall system performance is never reduced if physical layer performance is improved, which is the desired behaviour.

2.2 Utility Maximization Problems

The function u defined in Sect. 2.1 represents an abstraction of the upper layers: The parameter set $\mathcal{X}_2$, the mapping q and the constraint function $\boldsymbol{g}$ are condensed into the utility function u, whose value depends only on the rate vector that is provided by the physical layer.

Using the function u, the parameter optimization problem (2.14) can be equivalently written as

$$\max_{\boldsymbol{x}_1} u(\boldsymbol{r}(\boldsymbol{x}_1)) \quad \text{s.t.} \quad \boldsymbol{x}_1 \in \mathcal{X}_1. \tag{2.21}$$

The focus of this work is on using the degrees of freedom offered by advanced physical layer techniques to maximize overall system performance. From this perspective, the equivalent representation (2.21) is appealing: assuming that a function u is given, it allows to focus on the optimization of the physical layer parameters $\boldsymbol{x}_1$.

The representation (2.21), however, has one significant drawback: In general, the function u is not continuous on $\boldsymbol{r}(\mathcal{X}_1)$. This drawback is circumvented by introducing an additional constraint on $\boldsymbol{r}(\boldsymbol{x}_1)$ such that u is continuous on the subset of $\boldsymbol{r}(\mathcal{X}_1)$ where this constraint is fulfilled. To this end, first note that the constraint

$$\boldsymbol{g}(\boldsymbol{x}_2, \boldsymbol{s}) \geq \boldsymbol{0}$$

is equivalent to

$$\min_k g_k(\boldsymbol{x}_2, \boldsymbol{s}) \geq 0.$$

Define a function $h : \mathcal{D}_3 \to \mathbb{R}$ as follows:

$$h(\boldsymbol{s}) = \max_{\boldsymbol{x}_2 \in \mathcal{X}_2} \min_k g_k(\boldsymbol{x}_2, \boldsymbol{s}). \tag{2.22}$$

By Definition 2.1.1, $\boldsymbol{g}$ is continuous, which implies that $\boldsymbol{g}(\cdot, \boldsymbol{s})$ is continuous. By Corollary A.2.6, $\min_k g_k(\cdot, \boldsymbol{s})$ is continuous. Moreover, $\mathcal{X}_2$ is compact. By Proposition A.2.5, h is well-defined and continuous. In particular, for any $\boldsymbol{s} \in \mathcal{D}_3$, there exists an $\bar{\boldsymbol{x}}_2(\boldsymbol{s}) \in \mathcal{X}_2$ such that

$$h(\boldsymbol{s}) = \min_k g_k(\bar{\boldsymbol{x}}_2(\boldsymbol{s}), \boldsymbol{s}). \tag{2.23}$$

Corollary 2.2.1. *The function h is increasing.*

Proof. Let $\boldsymbol{s} \leq \boldsymbol{s}'$. From the monotonicity of $\boldsymbol{g}(\boldsymbol{x}_2, \cdot)$ and the definition of $\bar{\boldsymbol{x}}_2(\boldsymbol{s})$ it follows that

$$h(\boldsymbol{s}) = \min_k g_k(\bar{\boldsymbol{x}}_2(\boldsymbol{s}), \boldsymbol{s}) \le \min_k g_k(\bar{\boldsymbol{x}}_2(\boldsymbol{s}), \boldsymbol{s}') \le \max_{\boldsymbol{x}_2 \in \mathcal{X}_2} \min_k g_k(\boldsymbol{x}_2, \boldsymbol{s}') = h(\boldsymbol{s}'). \quad (2.24)$$

□

Proposition 2.2.2. *The set $\mathcal{F}_2(\boldsymbol{s})$, i.e., the set of feasible parameter setups at the upper layers for a given rate $\boldsymbol{s}$ provided by the physical layer, is nonempty if and only if*

$$h(\boldsymbol{s}) \ge 0. \quad (2.25)$$

Proof. First, assume that Inequality (2.25) holds. By (2.23), $\boldsymbol{g}(\bar{\boldsymbol{x}}_2(\boldsymbol{s}), \boldsymbol{s}) \ge \boldsymbol{0}$, and $\bar{\boldsymbol{x}}_2(\boldsymbol{s}) \in \mathcal{X}_2$. Thus, $\bar{\boldsymbol{x}}_2(\boldsymbol{s}) \in \mathcal{F}_2(\boldsymbol{s})$, i.e., $\mathcal{F}_2(\boldsymbol{s})$ is nonempty.

Second, assume that Inequality (2.25) does not hold. According to (2.22), this implies

$$\max_{\boldsymbol{x}_2 \in \mathcal{X}_2} \min_k g_k(\boldsymbol{x}_2, \boldsymbol{s}) < 0. \quad (2.26)$$

Equation (2.26) implies that there exists no $\boldsymbol{x}_2 \in \mathcal{X}_2$ such that $\boldsymbol{g}(\boldsymbol{x}_2, \boldsymbol{s}) \ge \boldsymbol{0}$, thus $\mathcal{F}_2(\boldsymbol{s})$ is empty. □

Corollary 2.2.3. *Any maximizer $\boldsymbol{x}_1^*$ of Problem (2.21) satisfies*

$$h(\boldsymbol{r}(\boldsymbol{x}_1^*)) \ge 0. \quad (2.27)$$

Define a set $\mathcal{R}_h$ as follows:

$$\mathcal{R}_h = \{\boldsymbol{s} \in \mathcal{D}_3 : h(\boldsymbol{s}) \ge 0\}. \quad (2.28)$$

Proposition 2.2.4. *The restriction of u to $\mathcal{R}_h$ is continuous.*

Proof. According to Proposition 2.2.2, the set $\mathcal{F}_2(\boldsymbol{s})$ is nonempty for any $\boldsymbol{s} \in \mathcal{R}_h$. By Lemma 2.1.1, $\mathcal{F}_2(\boldsymbol{s})$ is compact. Thus, for all $\boldsymbol{s} \in \mathcal{R}_h$,

$$u(\boldsymbol{s}) = \max_{\boldsymbol{x}_2 \in \mathcal{F}_2(\boldsymbol{s})} q(\boldsymbol{x}_2, \boldsymbol{s}). \quad (2.29)$$

The function q is continuous. Proposition A.2.5 yields the result. □

According to Corollary 2.2.3, Problem (2.21) can be written as

$$\max_{\boldsymbol{x}_1} u(\boldsymbol{r}(\boldsymbol{x}_1)) \quad \text{s.t.} \quad \boldsymbol{x}_1 \in \mathcal{X}_1, h(\boldsymbol{r}(\boldsymbol{x}_1)) \ge 0. \quad (2.30)$$

Let $\mathcal{F}_1$ denote the set of feasible physical layer parameter values:

$$\mathcal{F}_1 = \{\boldsymbol{x}_1 \in \mathcal{X}_1 : h(\boldsymbol{r}(\boldsymbol{x}_1)) \ge 0\}. \quad (2.31)$$

Proposition 2.2.5. *The set $\mathcal{F}_1$ is compact.*

Proof. The function $h \circ \boldsymbol{r}$ is continuous, hence the preimage of $\mathbb{R}_+$ under $h \circ \boldsymbol{r}$ is closed. By compactness of $\mathcal{X}_1$ and Proposition A.2.3,

$$\mathcal{F}_1 = \mathcal{X}_1 \cap (h \circ \boldsymbol{r})^{-1}(\mathbb{R}_+)$$

is compact. □

Recall from Definition 2.1.1 that $\boldsymbol{r}(\mathcal{X}_1) \subseteq \mathcal{D}_3$, thus

$$\boldsymbol{r}(\mathcal{F}_1) \subseteq \mathcal{R}_h.$$

As a result, Problem (2.30) corresponds to the maximization of a continuous function $u|_{\mathcal{R}_h} \circ \boldsymbol{r}|_{\mathcal{F}_1}$ over the compact set $\mathcal{F}_1$.

To summarize, continuity of q and $\boldsymbol{g}$ and compactness of $\mathcal{X}_2$ translate into continuity of the functions $u|_{\mathcal{R}_h}$ and h, while continuity of h and $\boldsymbol{r}$ and compactness of $\mathcal{X}_1$ result in compactness of $\mathcal{F}_1$. In addition, the monotonicity of the functions $q(\boldsymbol{y}, \cdot)$ and $\boldsymbol{g}(\boldsymbol{y}, \cdot)$ in the original parameter optimization problems results in increasing functions u and h. The last result is expected, as the pair of functions (u, h) characterizes the upper layers, and the monotonicity of u and h again guarantees that better performance at the physical layer does not lead to a performance degradation at upper layers.

Clearly, any problem of the type of Problem (2.30) with u, h, and $\boldsymbol{r}$ continuous, $\mathcal{X}_1$ compact, u and h increasing, and $\mathcal{F}_1$ nonempty is well-defined and obeys the monotonicity principle. This directly motivates the following definition:

Definition 2.2.1 (Utility Maximization Problem). Let $\mathcal{D}_1$ denote a metric space, let $\mathcal{X}_1 \subseteq \mathcal{D}_1$ be compact, and let $\boldsymbol{r} : \mathcal{D}_1 \to \mathbb{R}^K$ be a continuous function, for some K. Moreover, let $h : \mathcal{D}_h \to \mathbb{R}$ and $u : \mathcal{D}_u \to \mathbb{R}$ be continuous functions whose domains $\mathcal{D}_h$ and $\mathcal{D}_u$ are compatible in the following sense:

$$\boldsymbol{r}(\mathcal{X}_1) \subseteq \mathcal{D}_h \subseteq \mathbb{R}^K,$$
$$\{\boldsymbol{s} \in \mathcal{D}_h : h(\boldsymbol{s}) \geq 0\} \subseteq \mathcal{D}_u \subseteq \mathbb{R}^K.$$

Then

$$\max_{\boldsymbol{x}_1} u(\boldsymbol{r}(\boldsymbol{x}_1)) \quad \text{s.t.} \quad \boldsymbol{x}_1 \in \mathcal{X}_1, h(\boldsymbol{r}(\boldsymbol{x}_1)) \geq 0 \tag{2.32}$$

is a utility maximization problem if the following additional conditions hold:

(1) The set of feasible parameters is nonempty, i.e.,

$$\mathcal{F}_1 = \{\boldsymbol{x}_1 \in \mathcal{X}_1 : h(\boldsymbol{r}(\boldsymbol{x}_1)) \geq 0\} \neq \emptyset. \tag{2.33}$$

(2) The functions u and h are increasing.

A utility maximization problem can be conveniently represented by its data $(u, h, \boldsymbol{r}, \mathcal{X}_1)$. Accordingly, a tuple $(u, h, \boldsymbol{r}, \mathcal{X}_1)$ is said to be a utility maximization problem if it fulfills the conditions of Definition 2.2.1. For a given utility maximization problem $P = (u, h, \boldsymbol{r}, \mathcal{X}_1)$, the set of maximizers of Problem (2.32) is denoted by $\mathcal{X}^*(P)$. Hence, $\mathcal{X}^*(P) \subseteq \mathcal{X}_1$ contains the optimal parameter setups, and $u(\boldsymbol{r}(\mathcal{X}^*(P)))$ is the maximally achievable utility value.

A utility maximization problem consists of a *utility model* (u, h) and a physical layer model $(\boldsymbol{r}, \mathcal{X}_1)$. A utility model is said to be convex if both u and h are concave.[3] Likewise, a physical layer model is said to be convex if the rate map $\boldsymbol{r}$ is concave and the set of possible parameter setups is convex.[4]

The derivations of this section show that any parameter optimization problem corresponds to a utility maximization problem. The connection between both problems is established by (2.29) and (2.22). Moreover, with any utility maximization problem, a parameter optimization can be associated by letting

$$q(\boldsymbol{x}_2, \boldsymbol{s}) = u(\boldsymbol{s}), \tag{2.34}$$

$$g(\boldsymbol{x}_2, \boldsymbol{s}) = h(\boldsymbol{s}). \tag{2.35}$$

Obviously, this trivial solution corresponds to a model of the upper layers that has no parameters $\boldsymbol{x}_2$. It is not clear whether a nontrivial solution (i.e., a model of the upper layers that includes parameters) can be associated with an arbitrary utility model (u, h).

In the remainder of this work, utility maximization problems according to Definition 2.2.1 will be considered. Instead of specifying a detailed model of the upper layers in terms of a triple $(\mathcal{X}_2, q, \boldsymbol{g})$, a generic utility model in terms of a pair of functions (u, h) is used for the upper layers. In particular, the functions u and h are not specified explicitly, but are simply assumed to exhibit certain structural properties. Based on this abstract model of the upper layers, general strategies to find an optimum parameter setup at the physical layer are developed. The strategies are general in the sense that they are applicable to any utility model (u, h) that exhibits the necessary structure.

At this point, it is a valid question to ask why the more elaborate parameter optimization problem defined in Definition 2.1.1 was introduced, instead of starting right away with the utility maximization problem from Definition 2.2.1. The answer is twofold: First, a definition that includes parameters at the upper layers

[3]Note that concavity of u and h implies that their domains are convex sets, cf. Definition A.5.2. Moreover, concavity of h implies that the set

$$\{\boldsymbol{s} \in \mathcal{D}_h : h(\boldsymbol{s}) \geq 0\}$$

is convex.

[4]A prerequisite for defining concavity of $\boldsymbol{r}$ and convexity of $\mathcal{X}_1$ is that the domain of $\boldsymbol{r}$ is a subset of a real or complex vector space, cf. Definitions A.5.1 and A.5.2. Moreover, concavity of $\boldsymbol{r}$ implies that the domain of $\boldsymbol{r}$ is convex.

seems more natural. Second, deriving the utility function u from the less abstract model $(\mathcal{X}_2, q, \boldsymbol{g})$ shows which assumptions are implicitly made if certain structural properties of u are assumed. For example, a usual assumption on u is that u is concave. In light of (2.29), it becomes obvious that concavity is a rather strong assumption, even if the underlying functions $q(\boldsymbol{x}_2, \cdot)$ are concave: In general, the pointwise maximum of a family of concave functions is *not* concave. Similarly, the pointwise maximum of a family of smooth functions is *not* necessarily smooth.

2.3 Rate Region and Rate Space Problem

Throughout this section, let $P = (u, h, \boldsymbol{r}, \mathcal{X}_1)$ be a utility maximization problem according to Definition 2.2.1. The physical layer is specified by the pair $(\boldsymbol{r}, \mathcal{X}_1)$, where $\mathcal{X}_1$ is the set of possible physical layer parameter setups, and the function $\boldsymbol{r}$ assigns a vector $\boldsymbol{r}(\boldsymbol{x}_1)$ containing the values of K physical layer performance metrics to each parameter setup $\boldsymbol{x}_1 \in \mathcal{X}_1$. The physical layer performance metrics are assumed to express a "utility": if

$$\boldsymbol{r}(\boldsymbol{x}_1') \geq \boldsymbol{r}(\boldsymbol{x}_1),$$

then $\boldsymbol{x}_1'$ is said to provide better physical layer performance than $\boldsymbol{x}_1$. In other words, any performance metric for which a larger value is desirable can be used as a component function r_k of $\boldsymbol{r} = (r_1, \dots, r_K)$, such as information rate, signal-to-noise ratio (SNR), inverse mean-squared error (MSE), inverse delay, or inverse bit error rate (BER).

In the following, it is assumed that the k-th component function of $\boldsymbol{r}$ is the link rate of the k-th link in a wireless network having K links. Note that this is a terminological convention, and not a limitation that excludes alternative performance metrics. Preference for the term "rate" instead of the term "physical layer performance metric" is motivated as follows: First, the general term "physical layer performance metric" is rather cumbersome. Second, all physical layer models considered in Chap. 4 use information rate as performance metric. The solution methods presented in Chap. 3, however, only make assumptions on the structural properties of the map $\boldsymbol{r}$, but are indifferent to the precise nature of a performance metric, as long as it is consistent with the aforementioned assumptions.

Given a physical layer description $(\boldsymbol{r}, \mathcal{X}_1)$, the rate region is defined as the set of all rate vectors that can be provided by the system:

$$\mathcal{R} = \boldsymbol{r}(\mathcal{X}_1).$$

By Definition 2.2.1 and Proposition A.2.2, the rate region is a nonempty compact subset of $\mathbb{R}^K$. Based on the definition of the rate region, an equivalent formulation of a utility maximization is obtained:

$$\max_{s \in \mathcal{R}} u(s) \quad \text{s.t.} \quad h(s) \geq 0. \tag{2.36}$$

In the following, Problem (2.36) will be denoted as *rate space problem*. Define

$$\mathcal{H} = \{s \in \mathcal{D}_h : h(s) \geq 0\}.$$

By continuity of h, $\mathcal{H} = h^{-1}(\mathbb{R}_+)$ is closed. Moreover, Definition 2.2.1 implies that the set of feasible rates is nonempty:

$$\mathcal{R} \cap \mathcal{H} \neq \emptyset.$$

Intuitively, due to the monotonicity of u and h, utility is maximized by one of the largest rate vectors in $\mathcal{R}$. Due to the fact that for $K > 1$, "$\geq$" is a partial order, there usually does not exist a single "largest" vector in $\mathcal{R}$. Instead, the set of largest rate vectors is given by the set of Pareto optimal points par $(\mathcal{R})$. In other words, the maximum of the rate space problem is attained on the Pareto boundary of the rate region:

Proposition 2.3.1. *There exists* $s^* \in \text{par}(\mathcal{R})$ *such that* $h(s^*) \geq 0$ *and*

$$u(s^*) = \max_{s \in \mathcal{R}} u(s) \quad \text{s.t.} \quad h(s) \geq 0.$$

Proof. The rate region $\mathcal{R}$ is compact, $\mathcal{H}$ is closed, thus $\mathcal{R} \cap \mathcal{H}$ is compact. Moreover, u is continuous, and $\mathcal{R} \cap \mathcal{H}$ is nonempty, thus the maximum is attained in $\mathcal{R} \cap \mathcal{H}$. Let s' denote a maximizer. According to Proposition A.3.6, there exists $s^* \in \text{par}(\mathcal{R})$ such that $s^* \geq s'$. By monotonicity of u and h,

$$u(s^*) \geq u(s'),$$
$$h(s^*) \geq h(s') \geq 0,$$

thus s^* is also a maximizer. □

For a general nonempty compact rate region $\mathcal{R}$, little can be stated about its Pareto set par $(\mathcal{R})$, except for par $(\mathcal{R})$ being nonempty, cf. Proposition A.3.2. In particular, par $(\mathcal{R})$ is not necessarily closed, as shown by the following example.

Example 2.3.1. Let

$$\mathcal{R} = \left\{s \in \mathbb{R}_+^2 : \mathbf{1}^{\mathrm{T}} s \leq 1\right\} \cup \{(\alpha, 0) : 1 \leq \alpha \leq 2\}.$$

The set $\mathcal{R}$ is illustrated in Fig. 2.2. Then

$$\text{par}(\mathcal{R}) = \left(\left\{s \in \mathbb{R}_+^2 : \mathbf{1}^{\mathrm{T}} s = 1\right\} \setminus \{(1,0)\}\right) \cup \{(2,0)\}.$$

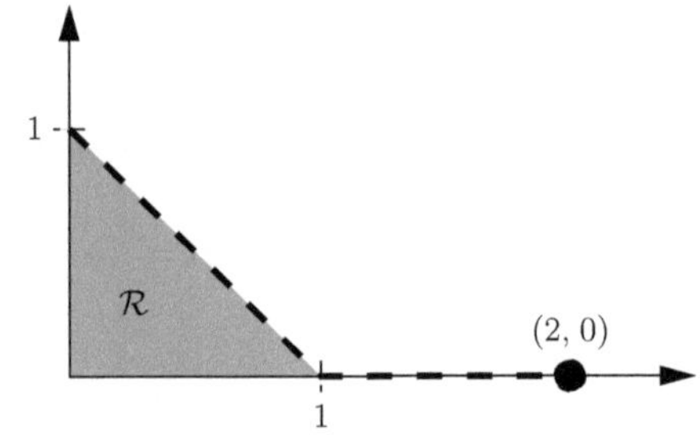

Fig. 2.2 The Pareto set of a closed set is not necessarily closed. The weak Pareto set is

While $\mathcal{R}$ is compact, par $(\mathcal{R})$ is not closed. Note that the weak Pareto set is indeed closed:

$$\operatorname{wpar}(\mathcal{R}) = \left\{ \boldsymbol{s} \in \mathbb{R}_+^2 : \mathbf{1}^{\mathrm{T}} \boldsymbol{s} = 1 \right\} \cup \left\{ (\alpha, 0) : 1 \le \alpha \le 2 \right\},$$

cf. the dotted line in Fig. 2.2. According to Proposition A.3.4, this is no coincidence.

While the Pareto set of a compact set may not be compact, the weak Pareto set inherits compactness from its original set, as shown by Proposition A.3.4. In other words, Proposition A.3.4 indicates that the weak Pareto set has a number of properties that are not necessarily shared by the Pareto set. This important difference is further exploited in Sect. 2.3.3.

2.3.1 Proper Rate Regions

The definition of a utility maximization problem implies that $\mathcal{R}$ is a nonempty compact subset of $\mathbb{R}^K$. In conjunction with the continuity of the functions u and h, these basic properties are sufficient for the existence of an optimal parameter setup. In Chap. 3, algorithms for computing an optimal parameter setup are discussed. For an algorithmic solution, the basic properties are generally not sufficient. Accordingly, for each of the solution methods discussed in Chap. 3, it is necessary to equip the rate region with some additional structure. In the following, three additional assumptions concerning the properties of $\mathcal{R}$ are made. All three assumptions have in common that they can be satisfied by formulating an equivalent utility optimization problem whose solution yields a maximizer of the original problem. In other words, the following three assumptions do not limit the applicability of the proposed framework. As demonstrated in Sects. 2.3.2 and 2.3.3, the three assumptions imply a number of properties of the set of weak Pareto optimal points. These properties are of fundamental importance for two of the solution methods discussed in Chap. 3.

Assumption 2.3.1. *The rate region* $\mathcal{R}$ *is a subset of* $\mathbb{R}_+^K$.

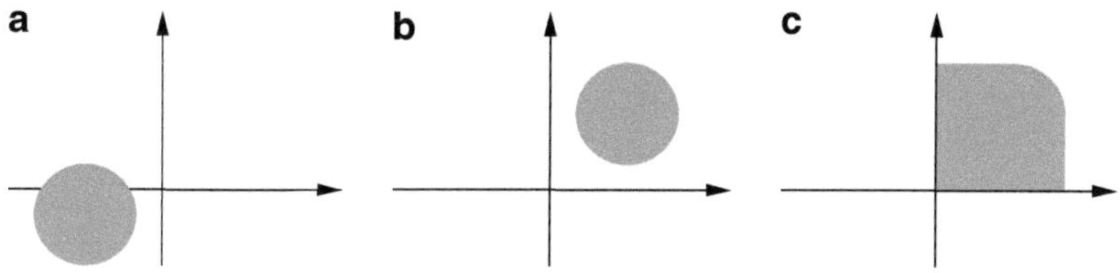

Fig. 2.3 From compact to comprehensive: Starting with an arbitrary compact set, a shift yields a compact subset of $\mathbb{R}_+^K$, and taking the comprehensive hull yields a comprehensive set

Given a utility maximization problem with rate region $\mathcal{R} \not\subset \mathbb{R}_+^K$, it is straightforward to define a utility maximization problem P' whose rate region $\mathcal{R}'$ is obtained by shifting $\mathcal{R}$ by $\boldsymbol{a} \in \mathbb{R}_+^K$ such that

$$\mathcal{R}' = \mathcal{R} + \{\boldsymbol{a}\} \in \mathbb{R}_+^K.$$

Due to the fact that $\mathcal{R}$ is a compact subset of $\mathbb{R}^K$, such a vector $\boldsymbol{a}$ always exists. The corresponding utility maximization problem is given by $P' = (u', h', \boldsymbol{r}', \mathcal{X}_1)$, with

$$u'(\boldsymbol{s}) = u(\boldsymbol{s} - \boldsymbol{a}),$$
$$h'(\boldsymbol{s}) = h(\boldsymbol{s} - \boldsymbol{a}),$$
$$\boldsymbol{r}'(\boldsymbol{x}_1) = \boldsymbol{r}(\boldsymbol{x}_1) + \boldsymbol{a}.$$

The problem P' is again a utility maximization problem according to Definition 2.2.1. Moreover, it has the same optimal parameter setups as the original problem P:

$$\mathcal{X}^*(P') = \mathcal{X}^*(P).$$

Assumption 2.3.2. *The rate region $\mathcal{R}$ is a comprehensive set.*

Any physical layer $(\boldsymbol{r}, \mathcal{X}_1)$ with $\mathcal{R} = \boldsymbol{r}(\mathcal{X}_1) \subseteq \mathbb{R}_+^K$ not comprehensive can be extended to a physical layer $(\boldsymbol{r}', \mathcal{X}_1')$ that has a comprehensive rate region.[5] The process of going from an arbitrary compact rate region to a comprehensive rate region is illustrated in Fig. 2.3.

In order to obtain a comprehensive rate region, the comprehensive hull operation is included in the definition of $(\boldsymbol{r}', \mathcal{X}_1')$: For any $\boldsymbol{s}' \in \text{comp}(\mathcal{R})$, there exists an $\boldsymbol{s} \in \mathcal{R}$ and a scaling vector $\boldsymbol{a} \in \mathbb{R}_+^K$ with $\boldsymbol{a} \leq \boldsymbol{1}$ such that $\boldsymbol{s}' = \boldsymbol{a}\boldsymbol{s}$. Based on this observation, define

[5]For a definition of comprehensiveness, see Appendix A.4.

$$\mathcal{A} = \{\boldsymbol{a} \in \mathbb{R}_+^K : \boldsymbol{a} \leq \mathbf{1}\}, \tag{2.37}$$

$$\mathcal{X}_1' = \mathcal{A} \times \mathcal{X}_1, \tag{2.38}$$

$$\boldsymbol{r}' : \mathcal{X}_1' \to \mathbb{R}^K, (\boldsymbol{a}, \boldsymbol{x}_1) \mapsto \boldsymbol{a}\boldsymbol{r}(\boldsymbol{x}_1). \tag{2.39}$$

Then

$$\mathcal{R}' = \boldsymbol{r}'(\mathcal{X}_1') = \operatorname{comp}(\mathcal{R}), \tag{2.40}$$

thus $\mathcal{R}'$ is comprehensive. Moreover, if P is a utility maximization problem, then $\mathcal{X}_1'$ is compact, $\boldsymbol{r}'$ is continuous, and $\mathcal{F}_1'$ is nonempty. As a result, $P' = (u, h, \boldsymbol{r}', \mathcal{X}_1')$ is a utility maximization problem. Finally,

$$(\boldsymbol{a}, \boldsymbol{x}_1) \in \mathcal{X}^*(P') \implies \boldsymbol{x}_1 \in \mathcal{X}^*(P), \tag{2.41}$$

as

$$\boldsymbol{a}\boldsymbol{r}(\boldsymbol{x}_1) \leq \boldsymbol{r}(\boldsymbol{x}_1),$$

thus

$$u(\boldsymbol{a}\boldsymbol{r}(\boldsymbol{x}_1)) \leq u(\boldsymbol{r}(\boldsymbol{x}_1)),$$
$$h(\boldsymbol{a}\boldsymbol{r}(\boldsymbol{x}_1)) \leq h(\boldsymbol{r}(\boldsymbol{x}_1)).$$

In other words, P can be solved by solving P'. Adding the comprehensive hull operation to the physical layer parameterization equips the rate region with the property of being comprehensive. The comprehensiveness of $\mathcal{R}$ plays an important role in the development of solution methods in the case that the rate region is not convex or (u, h) is not concave, cf. Chap. 3. However, as shown by (2.41), the physical layer does not have to implement the comprehensive hull operation, as utility is always maximized by a parameter setup with scaling $\boldsymbol{a} = \mathbf{1}$. This result is clearly not surprising, as the comprehensive hull only adds rate vectors that are smaller or equal (with respect to the partial order $\leq$ in $\mathbb{R}^K$), and smaller rate vectors cannot lead to better overall performance.

Note that a nonempty comprehensive rate region always contains the zero vector. A trivial nonempty, compact, and comprehensive rate region is given by $\mathcal{R} = \{\mathbf{0}\}$. Other examples are $\mathcal{R} = \operatorname{comp}(\{\boldsymbol{e}_1\})$ and

$$\mathcal{R} = \operatorname{comp}(\{(0, 1, 1), (1, 0, 1), (1, 1, 0)\}).$$

All three examples have in common that for any $\boldsymbol{s} \in \mathcal{R}$, the rate of at least one link is zero. Accordingly, the underlying utility maximization problems can be decomposed into a finite set of utility maximization problems with a reduced number of links. Utility maximization problems that have this property are excluded by assuming that there exists a rate vector where all links have non-zero rate, i.e., $\mathcal{R} \cap \mathbb{R}^K_{++} \neq \emptyset$.

Proposition 2.3.2. *A comprehensive subset $\mathcal{R}$ of $\mathbb{R}^K_+$ has nonempty interior if and only if*

$$\mathcal{R} \cap \mathbb{R}^K_{++} \neq \emptyset.$$

Proof. First, assume $\mathcal{R} \cap \mathbb{R}^K_{++} \neq \emptyset$. Let $s \in \mathcal{R} \cap \mathbb{R}^K_{++}$ and set

$$\delta = 0.5 \min_k s_k.$$

Note that $s \in \mathbb{R}^K_{++}$ implies $\delta > 0$. Then

$$\mathbf{0} \leq y \leq s, \forall y \in \mathcal{B}_\delta(0.5s),$$

which by comprehensiveness of $\mathcal{R}$ implies that $\mathcal{B}_\delta(0.5s) \subset \mathcal{R}$, cf. Fig. 2.4.

Second, assume that $\mathcal{R}$ has nonempty interior. Accordingly, there exists a pair (s, δ) such that $\mathcal{B}_\delta(s) \subset \mathcal{R}$. The ball $\mathcal{B}_\delta(s)$ is open, hence there exists $\epsilon > 0$ such that $s + \epsilon\mathbf{1} \in \mathcal{B}_\delta(s)$. But $s \in \mathbb{R}^K_+$ and $\epsilon\mathbf{1} \in \mathbb{R}^K_{++}$, thus $s + \epsilon\mathbf{1} \in \mathbb{R}^K_{++}$. □

Assumption 2.3.3. *The rate region $\mathcal{R}$ has nonempty interior.*

As for the first assumption, Assumption 2.3.3 can be fulfilled by shifting the rate region. In other words, a utility maximization P that does not satisfy Assumption 2.3.3 can be solved by solving a shifted version P'.

In the following, the term "proper" is used for a rate region that satisfies the aforementioned three assumptions:

Definition 2.3.1 (Proper Set). A nonempty, comprehensive, compact subset of $\mathbb{R}^K_+$ that has nonempty interior is denoted as *proper*.

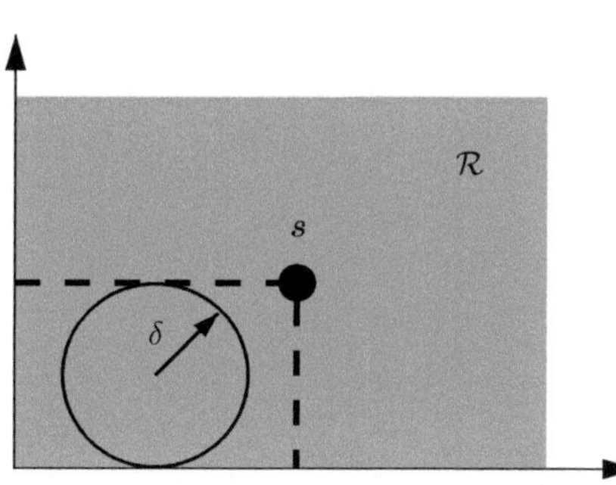

Fig. 2.4 Existence of a point $s \in \mathcal{R} \cap \mathbb{R}^K_{++}$ implies that comprehensive set $\mathcal{R}$ has nonempty interior

2.3.2 Intersection Problems

Based on the observation that utility is maximized by a rate vector in par $(\mathcal{R})$, it is of interest to characterize the set of Pareto optimal points. However, in Sect. 2.3.1 it was already indicated that, in general, the Pareto set has little structure. In contrast, the weak Pareto set of a proper rate region offers a rather rich structure, as discussed in Sect. 2.3.3. A first step in the characterization of wpar $(\mathcal{R})$ is the provision of a basic method for computing points in wpar $(\mathcal{R})$, which is the subject of this section.

Let $\phi : \mathbb{R} \to \mathbb{R}^K$ be a continuous function. Being a continuous function from $\mathbb{R}$ to $\mathbb{R}^K$, ϕ is a curve in $\mathbb{R}^K$. Let $\mathcal{R}$ be a proper rate region and assume

$$\exists \underline{\gamma} \in \mathbb{R} : \phi(\underline{\gamma}) \in \mathcal{R}. \tag{2.42}$$

Then, starting at $\gamma = \underline{\gamma}$, increasing γ leads to a movement of $\phi(\gamma)$ in $\mathcal{R}$ (assuming that ϕ is not constant). In addition, if there exists $\overline{\gamma}$ such that ϕ restricted to $\{\overline{\gamma}\}+\mathbb{R}_+$ is strictly increasing, i.e.,

$$\exists \overline{\gamma} \in \mathbb{R} : \gamma' > \gamma \geq \overline{\gamma} \implies \phi(\gamma') > \phi(\gamma), \tag{2.43}$$

then boundedness of $\mathcal{R}$ implies that there exists γ' such that $\phi(\gamma') \notin \mathcal{R}$. Moreover, due to the fact that $\mathcal{R}$ is closed and ϕ is continuous, there exists a point where the image of ϕ intersects the boundary of $\mathcal{R}$. Such a point can be computed by solving the line-search

$$\max \{\gamma \in \mathbb{R} : \phi(\gamma) \in \mathcal{R}\}. \tag{2.44}$$

This intuitive result is illustrated in Fig. 2.5 and proven in Proposition A.3.5.

Definition 2.3.2 (Intersection Problem). Let $\mathcal{R}$ be a nonempty compact subset of $\mathbb{R}_+^K$ and $\phi : \mathbb{R} \to \mathbb{R}^K$ a continuous function satisfying Conditions (2.42) and (2.43). The problem of determining $\phi(\gamma^*)$, where

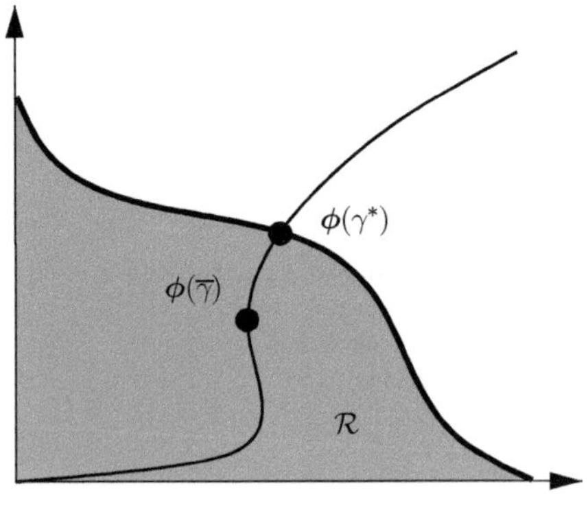

Fig. 2.5 The solution of an intersection problem is weak Pareto optimal

$$\gamma^* = \max\{\gamma \in \mathbb{R} : \phi(\gamma) \in \mathcal{R}\},$$

is denoted as *intersection problem.*

Note that despite the fact that Problem (2.44) involves a one-dimensional line search in γ, solving the intersection problem may be arbitrarily hard. If $\phi(\overline{\gamma}) \in \mathcal{R}$, then

$$\{\gamma \in \mathbb{R} : \gamma \geq \overline{\gamma}, \phi(\gamma) \in \mathcal{R}\}$$

is convex, due to the fact that ϕ is strictly increasing for $\gamma \geq \overline{\gamma}$ and $\mathcal{R}$ is comprehensive. In this case, an efficient solution is possible if a computationally efficient membership test for the rate region $\mathcal{R}$ is available.

From the viewpoint of characterizing the weak Pareto boundary of $\mathcal{R}$, an obvious question to ask is under what conditions the intersection point $\phi(\gamma^*)$ is weak Pareto optimal. The answer is provided by Proposition A.4.4: If, in addition to being nonempty and compact, $\mathcal{R}$ is comprehensive, then $\phi(\gamma^*) \in \operatorname{wpar}(\mathcal{R})$. In other words, for a proper rate region, the solution of any intersection problem is weak Pareto optimal.

Throughout this work, two definitions of a function $\phi : \mathbb{R} \to \mathbb{R}^K$ satisfying Conditions (2.42) and (2.43) are frequently used:

(1) $\phi(\gamma) = \gamma \boldsymbol{v}$, with $\boldsymbol{v} \in \mathbb{R}^K_+$, and
(2) $\phi(\gamma) = (\boldsymbol{s} + \gamma \boldsymbol{d})^+$, with $\boldsymbol{s} \in \mathbb{R}^K_+$ and $\boldsymbol{d} \in \mathbb{R}^K_{++}$.

A third possible choice of ϕ is given by

$$\phi(\gamma) = \boldsymbol{s} + \gamma \boldsymbol{d},$$

with $\boldsymbol{s} \in \mathcal{R}$ and $\boldsymbol{d} \in \mathbb{R}^K_+$. This choice of ϕ together with Proposition A.4.4 yields the following result: Starting at any point in $\mathcal{R}$ and moving in any positive direction $\boldsymbol{d}$, the trajectory will eventually intersect with the weak Pareto set—in other words, the weak Pareto set of a proper rate region has no holes.

Corollary 2.3.3. *Let $\mathcal{R}$ be a proper set. Then* $\operatorname{wpar}(\mathcal{R})$ *is connected.*

Example 2.3.1 shows that the Pareto set of a proper rate region does not share this property in general.

2.3.3 The Pareto Manifold

Based on the assumption that the rate region is proper, two topological properties of its weak Pareto set have already been established, namely that $\operatorname{wpar}(\mathcal{R})$ is compact and connected. Section 2.3.2 shows that intersection problems are a basic method for computing points in the weak Pareto set of a proper rate region. In the next step, this basic method is used to make further statements about the topological properties of the weak Pareto set of a proper rate region.

In this section, it is shown that the weak Pareto set of a proper rate region is essentially a topological manifold of dimension $K-1$.[6] The result is established by constructing a homeomorphism between an open subset of $\mathbb{R}^{K-1}$ and $\mathrm{wpar}(\mathcal{R}) \cap \mathbb{R}^K_{++}$.[7]

Let a set $\mathcal{S}$ be defined as follows:

$$\mathcal{S} = \left\{ \boldsymbol{v} \in \mathbb{R}^K_{++} : \mathbf{1}^{\mathrm{T}} \boldsymbol{v} = 1 \right\}.$$

Moreover, choose a matrix $\boldsymbol{Q} \in \mathbb{R}^{K \times K-1}$ such that the columns of $\boldsymbol{Q}$ constitute a basis of $\mathrm{null}\left(\mathbf{1}^{\mathrm{T}}\right)$:

$$\mathrm{range}\,(\boldsymbol{Q}) = \mathrm{null}\left(\mathbf{1}^{\mathrm{T}}\right).$$

Given $\boldsymbol{Q}$, define a set $\mathcal{X}$ and a map $\phi : \mathcal{X} \to \mathcal{S}$ as follows:

$$\mathcal{X} = \left\{ \boldsymbol{x} \in \mathbb{R}^{K-1} : \frac{1}{K}\mathbf{1} + \boldsymbol{Q}\boldsymbol{x} \in \mathbb{R}^K_{++} \right\},$$

$$\phi(\boldsymbol{x}) = \frac{1}{K}\mathbf{1} + \boldsymbol{Q}\boldsymbol{x}.$$

Note that $\mathcal{X}$ is an open subset of $\mathbb{R}^{K-1}$, as it is the preimage of an open set under a continuous function.

Lemma 2.3.4. *The map ϕ is a homeomorphism.*

Proof. The inverse map $\phi^{-1} : \mathcal{S} \to \mathcal{X}$ is given by

$$\phi^{-1}(\boldsymbol{v}) = (\boldsymbol{Q}^{\mathrm{T}}\boldsymbol{Q})^{-1}\boldsymbol{Q}^{\mathrm{T}}\boldsymbol{v}.$$

Both ϕ and ϕ^{-1} are continuous. □

Given a proper rate region $\mathcal{R}$, define a map $\alpha : \mathcal{S} \to \mathbb{R}$,

$$\alpha(\boldsymbol{v}) = \max\left\{x \in \mathbb{R} : x\boldsymbol{v} \in \mathcal{R}\right\}.$$

By Proposition A.3.5, the value $\alpha(\boldsymbol{v})$ is well-defined for all $\boldsymbol{v} \in \mathcal{S}$ and any proper set $\mathcal{R}$.

Lemma 2.3.5. *Let $\mathcal{R}$ be a proper set and let $\boldsymbol{v} \in \mathcal{S}$. Then*

[6] A definition of the term *topological manifold* is provided in Appendix A.8.

[7] A *homeomorphism* is a continuous function that has a continuous inverse function, cf. Definition A.2.2 in Appendix A.2.

Fig. 2.6 Illustration of Lemma 2.3.5

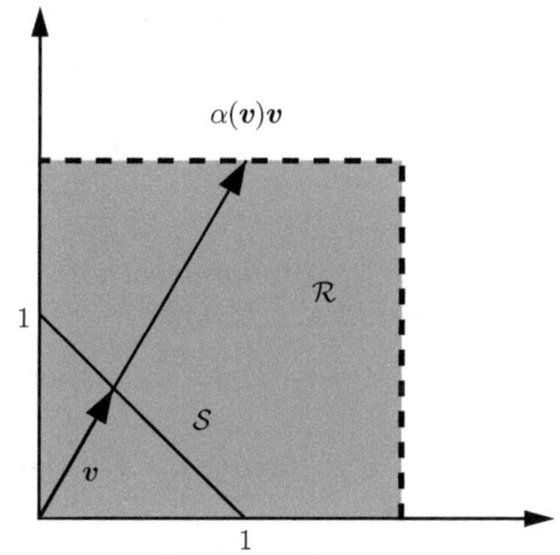

$$\alpha(v)v \in \operatorname{wpar}(\mathcal{R}) \cap \mathbb{R}^K_{++}.$$

Proof. The property $\alpha(v)v \in \operatorname{wpar}(\mathcal{R})$ follows from Proposition A.4.4. The rate region $\mathcal{R}$ is proper, thus there exists $s \in \mathcal{R}$ such that $s \gg \mathbf{0}$. Accordingly, by Proposition A.4.3, $\alpha(v) > 0$. Finally, $\mathcal{S} \subset \mathbb{R}^K_{++}$, thus $\alpha(v)v \in \mathbb{R}^K_{++}, \forall v \in \mathcal{S}$. □

Lemma 2.3.5 states that the point $\alpha(v)v$ is weak Pareto optimal and strictly positive. The strict positivity of $\alpha(v)v$ is needed in the proof of Lemma 2.3.6.

Example 2.3.2. Lemma 2.3.5 is illustrated in Fig. 2.6. In the example, the rate region $\mathcal{R}$ is the comprehensive hull of the point $(2, 2)$. The weak Pareto set of $\mathcal{R}$ is shown as a dashed line. Note that in the example, the Pareto set is a strict subset of the weak Pareto set. In fact, the Pareto set has only one element, the point $(2, 2)$. As a result, there is only one $v \in \mathcal{S}$ such that $\alpha(v)v \in \operatorname{par}(\mathcal{R})$, which shows that $\operatorname{wpar}(\mathcal{R})$ cannot be replaced by $\operatorname{par}(\mathcal{R})$ in Lemma 2.3.5.

Example 2.3.3. Let $\mathcal{R} = \operatorname{comp}(\{(1, 0), (0, 1)\})$. The set $\mathcal{R}$ is nonempty, compact, and comprehensive, but not proper. The map α is well-defined for any nonempty compact and comprehensive set. However, for this choice of $\mathcal{R}$, $\alpha(v) = 0, \forall v \in \mathcal{S}$. Obviously, for $\alpha(v)v$ to be strictly positive, comprehensiveness is not sufficient.

Based on Fig. 2.6, it can be conjectured that by computing $\alpha(v)v$ for all v in $\mathcal{S}$, it is possible to obtain all strictly positive points in the weak Pareto set of $\mathcal{R}$. Likewise, given a point in $\operatorname{wpar}(\mathcal{R})$, by proper scaling, a point in $\mathcal{S}$ is obtained. Based on this observation, Lemma 2.3.6 establishes a topological equivalence between $\operatorname{wpar}(\mathcal{R}) \cap \mathbb{R}^K_{++}$ and $\mathcal{S}$.

Lemma 2.3.6. *Let $\mathcal{R}$ be a proper set. Define a map $\psi : \mathcal{S} \to \operatorname{wpar}(\mathcal{R}) \cap \mathbb{R}^K_{++}$ as follows:*

$$\psi(v) = \alpha(v)v.$$

The map ψ is a homeomorphism.

Proof. The map ψ is surjective, i.e., $\psi(\mathcal{S}) = \text{wpar}(\mathcal{R}) \cap \mathbb{R}^K_{++}$: Assume there exists $s \in \text{wpar}(\mathcal{R}) \cap \mathbb{R}^K_{++}$ with $s \notin \psi(\mathcal{S})$. Let

$$v = \frac{s}{\mathbf{1}^{\mathrm{T}} s}.$$

From $s \in \mathbb{R}^K_{++}$ it follows that $v \in \mathcal{S} \subset \mathbb{R}^K_{++}$. The assumption $s \notin \psi(\mathcal{S})$ implies that there exists $s' \in \mathcal{R}$ with

$$s' = \alpha(v)v$$

and $\alpha(v) > \mathbf{1}^{\mathrm{T}} s$. But $\alpha(v) > \mathbf{1}^{\mathrm{T}} s$ and $v \in \mathbb{R}^K_{++}$ implies $s' \gg s$, which contradicts $s \in \text{wpar}(\mathcal{R})$.

The inverse map $\psi^{-1} : \text{wpar}(\mathcal{R}) \cap \mathbb{R}^K_{++} \to \mathcal{S}$ is given by

$$\psi^{-1}(x) = \frac{x}{\mathbf{1}^{\mathrm{T}} x}.$$

Moreover, ψ^{-1} is continuous. To show that ψ is continuous, it is sufficient to show that the map α is continuous. Let $v, w \in \mathcal{S}$. From $\alpha(v)v, \alpha(w)w \in \text{wpar}(\mathcal{R}) \cap \mathbb{R}^K_{++}$ it follows that for any $v, w \in \mathcal{S}$,

$$\frac{\alpha(v)}{\alpha(w)} \min_k \frac{v_k}{w_k} \le 1 \le \frac{\alpha(v)}{\alpha(w)} \max_k \frac{v_k}{w_k}. \tag{2.45}$$

Re-arranging (2.45) yields

$$\alpha(v) \min_k \frac{v_k}{w_k} \le \alpha(w) \le \alpha(v) \max_k \frac{v_k}{w_k}.$$

Let $\{v^{(n)}\}$ be a sequence in $\mathcal{S}$ such that $\lim_{n\to\infty} v^{(n)} = v$. Then by continuity,

$$\lim_{n\to\infty} \min_k \frac{v_k}{v_k^{(n)}} = 1,$$

$$\lim_{n\to\infty} \max_k \frac{v_k}{v_k^{(n)}} = 1.$$

Thus,

$$\alpha(v) \le \lim_{n\to\infty} \alpha(v^{(n)}) \le \alpha(v),$$

implying that

$$\lim_{n\to\infty} \alpha(v^{(n)}) = \alpha(v),$$

which shows the continuity of α. □

Lemma 2.3.6 establishes a topological equivalence between the sets $\text{wpar}(\mathcal{R}) \cap \mathbb{R}^K_{++}$ and $\mathcal{S}$. Lemma 2.3.4 already established an equivalence between an open subset of $\mathbb{R}^{K-1}_+$ and $\mathcal{S}$. Based on these results, we can make the following fundamental statement about the weak Pareto set of a proper set:

Proposition 2.3.7. *Let $\mathcal{R}$ be a proper set. Then* $\text{wpar}(\mathcal{R}) \cap \mathbb{R}^K_{++}$ *is a topological manifold of dimension* $K-1$.

Proof. By Lemmas 2.3.4 and 2.3.6, the map $\psi \circ \phi : \mathcal{X} \to \text{wpar}(\mathcal{R}) \cap \mathbb{R}^K_{++}$ is a homeomorphism. Moreover, $\mathcal{X}$ is an open subset of $\mathbb{R}^{K-1}$. □

Definition 2.3.3 (Pareto Manifold). Let $\mathcal{R}$ be a proper set. Then $\text{wpar}(\mathcal{R}) \cap \mathbb{R}^K_{++}$ is denoted as *Pareto manifold*.

By Proposition 2.3.1, $\text{wpar}(\mathcal{R})$ contains a maximizer of the rate space problem. Proposition 2.3.7 shows that $\text{wpar}(\mathcal{R}) \cap \mathbb{R}^K_{++}$ is a manifold. Based on these results, Sect. 3.5 develops a local method that operates in the Pareto manifold.

2.3.4 Outer-Approximation of a Proper Rate Region

Two of the solution approaches presented in Chap. 3 require the outer approximation of a rate region. If the rate region is proper, a simple rectangular outer approximation can be obtained from the so-called single-link points of the rate region, as shown in the following.

Definition 2.3.4. Let $\mathcal{R}$ be a nonempty compact subset of $\mathbb{R}^K_+$. A point $s' \in \mathcal{R}$ is a *single-link point* of $\mathcal{R}$ if $s' \neq \mathbf{0}$ and there exists $k \in \{1, \dots, K\}$ such that

$$s' = \gamma_k^* e_k,$$

where

$$\gamma_k^* = \max\{\gamma \in \mathbb{R}_+ : \gamma e_k \in \mathcal{R}\}.$$

Note that by Proposition A.3.5, γ_k^* is well defined for all k. A single-link point is a point where the rate of one link is positive, while the rates of all other links are zero. Single-link points are interesting from a computational perspective, as it is often easier to solve a single-link problem than a multi-link problem.

Clearly, not every set has single-link points.[8] Moreover, a subset of $\mathbb{R}^K_+$ can have up to K single-link points, one for each dimension.

Corollary 2.3.8. *A proper rate region $\mathcal{R}$ has K single-link points.*

---

[8]As a simple example, any subset of $\mathbb{R}^K_{++}$ has no single-link points.

Definition 2.3.5. Let $\mathcal{R}$ be a nonempty compact subset of $\mathbb{R}_{+}^{K}$. The utopia point of $\mathcal{R}$ is defined as follows:

$$\boldsymbol{s}^{\mathrm{U}} \text{ is the utopia point of } \mathcal{R}. \iff s_k^{\mathrm{U}} = \max_{\boldsymbol{s} \in \mathcal{R}} s_k, \forall k \in \{1, \ldots, K\}. \tag{2.46}$$

The utopia point is not achievable if the maximal rate on one link can only be achieved by reducing the rate on other links. In a wireless system, this is usually the case, as links are coupled via resource constraints and interference. Conversely, if the utopia point is achievable, no trade-off between links is required. This indicates that links are completely decoupled and can be treated separately.

Example 2.3.4. Single-link points and utopia point of a proper rate region $\mathcal{R}$ are illustrated in Fig. 2.7. In Fig. 2.7, the single-link points are $\boldsymbol{s}^1 = (2, 0)$ and $\boldsymbol{s}^2 = (0, 2)$. The utopia point is $\boldsymbol{s}^{\mathrm{U}} = (2, 2)$. Evidently, the utopia point is not achievable. In contrast, Fig. 2.6 shows a rate region where the utopia point actually is achievable.

Corollary 2.3.9. *Let $\boldsymbol{s}^U$ denote the utopia point of a rate region $\mathcal{R}$. Then $\mathcal{R}$ is contained in the hyperrectangle $[\mathbf{0}, \boldsymbol{s}^U]$, i.e.,*

$$\mathcal{R} \subseteq [\mathbf{0}, \boldsymbol{s}^U].$$

Proof. By assumption, $\mathcal{R} \subset \mathbb{R}_{+}^{K}$. Let $\boldsymbol{s}' \in \mathcal{R}$. Thus,

$$0 \le s_k' \le \max_{\boldsymbol{s} \in \mathcal{R}} s_k, \forall k \implies \mathbf{0} \le \boldsymbol{s}' \le \boldsymbol{s}^{\mathrm{U}} \implies \boldsymbol{s}' \in [\mathbf{0}, \boldsymbol{s}^{\mathrm{U}}] \implies \mathcal{R} \subseteq [\mathbf{0}, \boldsymbol{s}^{\mathrm{U}}].$$

□

The hyperrectangle $[\mathbf{0}, \boldsymbol{s}^{\mathrm{U}}]$ provides a simple outer-approximation of the rate region $\mathcal{R}$. The existence of such an outer-approximation follows from the compactness

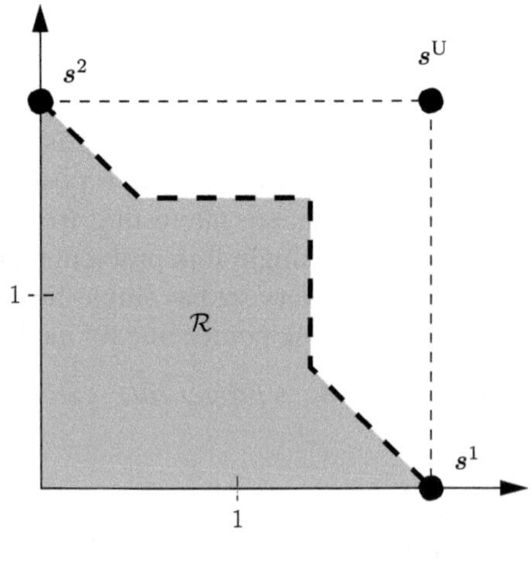

Fig. 2.7 Single-link points and utopia point

of $\mathcal{R}$. A proper rate region $\mathcal{R}$ is also comprehensive. Comprehensiveness yields a relation between the utopia point and the single-link points of $\mathcal{R}$, cf. Fig. 2.7.

Proposition 2.3.10. *Let $\mathcal{R}$ be a proper rate region. Let $\boldsymbol{s}^U$ denote the utopia point of $\mathcal{R}$. Moreover, let $\{\boldsymbol{s}^1, \ldots, \boldsymbol{s}^K\}$ denote the set of single-link point of $\mathcal{R}$. Then*

$$\boldsymbol{s}^U = \sum_{k=1}^{K} \boldsymbol{s}^k. \tag{2.47}$$

Proof. By definition of a proper rate region, $\mathcal{R}$ is nonempty and compact, thus $\boldsymbol{s}^{\mathrm{U}}$ is well-defined. Moreover, a proper rate region has K single-link points. Note that (2.47) is equivalent to

$$s_k^{\mathrm{U}} = s_k^k, \forall k.$$

From $\boldsymbol{s}^k \in \mathcal{R}$ and the definition of $\boldsymbol{s}^{\mathrm{U}}$ it follows that

$$s_k^{\mathrm{U}} \geq s_k^k. \tag{2.48}$$

Moreover, by definition of $\boldsymbol{s}^{\mathrm{U}}$, there exists $\boldsymbol{s}' \in \mathcal{R}$ with $s'_k = s_k^{\mathrm{U}}$. From $\mathcal{R} \subset \mathbb{R}_+^K$ it follows that $\mathbf{0} \leq s'_k \boldsymbol{e}_k \leq \boldsymbol{s}'$. A proper rate region $\mathcal{R}$ is comprehensive, thus $s'_k \boldsymbol{e}_k \in \mathcal{R}$. By definition of s_k^k,

$$s_k^k \geq s'_k = s_k^{\mathrm{U}}. \tag{2.49}$$

Combining (2.48) and (2.49) yields the result. □

2.3.5 Convex Rate Regions

The rate region $\mathcal{R}$ is not necessarily convex. In Sect. 2.3.1, it is shown that a comprehensive rate region can be obtained by including the comprehensive hull operation in the physical layer parameterization. Similarly, a convex rate region can be obtained by extending the physical layer parameterization $(\boldsymbol{r}, \mathcal{X}_1)$ such that it includes a convex hull operation. In other words, a convex rate region

$$\mathcal{C} = \operatorname{conv}(\mathcal{R}) \tag{2.50}$$

can be associated with each (nonconvex) rate region $\mathcal{R}$.

Proposition 2.3.11. *Let $\mathcal{R} \subset \mathbb{R}_+^K$ be nonempty, compact, and comprehensive. Moreover, let $\mathcal{R}$ have K single-link points. Then $\mathcal{C} = \operatorname{conv}(\mathcal{R})$ is proper.*

Proof. Nonemptyness is obvious. By Theorem A.5.3 $\mathcal{C}$ is compact, by Proposition A.5.4 $\mathcal{C}$ is comprehensive. It remains to show that $\mathcal{C} \cap \mathbb{R}_{++}^K \neq \emptyset$. Let $\{s^1, \ldots, s^K\}$ denote the set of single-link points of $\mathcal{R}$. By Definition 2.3.4, $s^k = \gamma_k^* e_k$, with $\gamma_k^* > 0$. Define

$$s' = \frac{1}{K} \sum_{k=1}^{K} s^k.$$

Then $s' \gg \mathbf{0}$ and, by convexity of $\mathcal{C}$, $s' \in \mathcal{C}$. □

Corollary 2.3.12. *The convex hull of a proper set is proper.*

According to Carathéodory's Theorem [22], each point in $\mathcal{C}$ can be written as a convex combination of at most $K + 1$ points in $\mathcal{R}$. The maximum number of necessary points can be further reduced if $\mathcal{R}$ compact and comprehensive, as shown by Proposition 2.3.13.

Proposition 2.3.13. *Let $\mathcal{R} \subset \mathbb{R}_+^K$ be compact and comprehensive. Each point in $\mathcal{C} = \operatorname{conv}(\mathcal{R})$ can be written as a convex combination of at most K points in $\mathcal{R}$.*

Proof. If $\mathcal{R}$ is empty, the result is obvious. Assume that $\mathcal{R}$ is nonempty. Let $s' \in \mathcal{C}$. If $s' = \mathbf{0}$, the result is again obvious, as $\mathbf{0} \in \mathcal{R}$ for any comprehensive set $\mathcal{R}$. In the following, let $s' \in \mathcal{C} \setminus \{\mathbf{0}\}$. According to Proposition A.4.4, for each $s' \in \mathcal{C} \setminus \{\mathbf{0}\}$ there exists a scaling γ and a rate vector $s \in \operatorname{wpar}(\mathcal{C})$ such that

$$\gamma s' = s.$$

Moreover, from $s' \in \mathcal{C}$ and $s \in \operatorname{wpar}(\mathcal{C})$ it follows that $\gamma \geq 1$. By Proposition A.3.1, $\operatorname{wpar}(\mathcal{C}) \subseteq \operatorname{bd}(\mathcal{C})$. Hence, by Proposition A.5.7, each point in $\operatorname{wpar}(\mathcal{C})$ can be written as a convex combination of at most K points in $\mathcal{R}$, i.e., for any $s \in \operatorname{wpar}(\mathcal{C})$ there exist coefficients $b_k \geq 0$ and points $s^k \in \mathcal{R}$ such that $\sum_{k=1}^K b_k = 1$ and

$$s = \sum_{k=1}^{K} b_k s^k.$$

Consequently,

$$s' = \frac{1}{\gamma} \sum_{k=1}^{K} b_k s^k = \sum_{k=1}^{K} b_k \frac{1}{\gamma} s^k.$$

Due to the fact that $\mathcal{R}$ is comprehensive and $\gamma^{-1} \leq 1$, $\gamma^{-1} s^k \in \mathcal{R}$. As a result, $s' \in \mathcal{C}$ can be written as a convex combination of at most K points in $\mathcal{R}$. □

Based on Proposition 2.3.13, define

$$\mathcal{B} = \left\{ \boldsymbol{b} \in \mathbb{R}_+^K : \mathbf{1}^{\mathrm{T}}\boldsymbol{b} = 1 \right\},$$
$$\mathcal{X}_1' = \mathcal{B} \times \mathcal{X}_1^K,$$
$$\boldsymbol{r}' : \mathcal{X}_1' \to \mathbb{R}^K, (\boldsymbol{b}, \boldsymbol{x}_1^1, \ldots, \boldsymbol{x}_1^K) \mapsto \sum_{k=1}^{K} b_k \boldsymbol{r}(\boldsymbol{x}_1^k). \tag{2.51}$$

Then

$$\mathcal{C} = \boldsymbol{r}'(\mathcal{X}_1')$$

is the convex hull of $\mathcal{R} = \boldsymbol{r}(\mathcal{X}_1)$. Moreover, if $(u, h, \boldsymbol{r}, \mathcal{X}_1)$ is a utility maximization problem, then $(u, h, \boldsymbol{r}', \mathcal{X}_1')$ is a utility maximization problem.

It is important to note that even though $\boldsymbol{r}'(\mathcal{X}_1')$ is convex, the function $\boldsymbol{r}'$ is nonconcave if $\boldsymbol{r}$ is nonconcave[9]—in other words, including the convex hull operation in the parameterization does not remove the nonconcavity with respect to the physical layer parameters. Still, at the rate level, the physical layer $(\boldsymbol{r}', \mathcal{X}')$ is characterized by a convex rate region $\mathcal{C}$. This property is exploited by the so-called rate space methods presented in Chap. 3.

Including the convex hull operation increases the number of system parameters. If there exists $k \in \{1, \ldots, K\}$ such that

$$(\boldsymbol{e}_k, \boldsymbol{x}_1^1, \ldots, \boldsymbol{x}_1^K) \in \mathcal{X}^*(P'),$$

for some $(\boldsymbol{x}_1^1, \ldots, \boldsymbol{x}_1^K) \in \mathcal{X}_1^K$, then the convex hull operation is not required, as the optimum rate vector in $\mathcal{C}$ is also in $\mathcal{R}$. In general, though, the optimal parameter setup of problem P' requires the convex hull operation, and the maximum utility value of problem P' may be larger than the maximum utility value of P. As a result, if the original rate region $\mathcal{R}$ is nonconvex, and the convex rate region $\mathcal{C}$ is used to determine an optimal parameter setup, the physical layer has to implement the convex hull operation.

If information rates are used as performance metrics, one possible way to implement the convex hull operation is *time sharing* between the parameter setups $\boldsymbol{x}_1^1, \ldots, \boldsymbol{x}_1^K$, with time sharing coefficients b_k. Time-sharing can also be interpreted as scheduling with variable slot lengths, where the scheduler chooses each of the K rate vectors $\boldsymbol{r}(\boldsymbol{x}_1^1), \ldots, \boldsymbol{r}(\boldsymbol{x}_1^K)$ for a certain fraction of time.

---

[9]The statement follows by letting $\boldsymbol{x}_1' = (\mathrm{e}_1, \boldsymbol{x}_1, \ldots, \boldsymbol{x}_1)$. Then $\boldsymbol{r}'(\boldsymbol{x}_1') = \boldsymbol{r}(\boldsymbol{x}_1)$. Thus, if $\boldsymbol{r}$ is nonconcave, there exist $\boldsymbol{y}, \boldsymbol{z} \in \mathcal{X}_1$ such that $\boldsymbol{r}'$ violates the concavity condition on the line segment $\{\alpha(\mathrm{e}_1, \boldsymbol{y}) + (1-\alpha)(\mathrm{e}_1, \boldsymbol{z}), \alpha \in [0, 1]\} \subset \mathcal{X}'$.

Chapter 3
Solution Methods

In Chap. 2, a general model of an utility maximization problem is defined, cf. Definition 2.2.1. The data of the problem is specified in terms of a tuple $(u, h, \boldsymbol{r}, \mathcal{X}_1)$. In this chapter, different methods to *solve* a utility maximization problem are presented. The applicability of each method depends on the properties of the elements u, h, $\boldsymbol{r}$, and $\mathcal{X}_1$. Definition 2.2.1 already postulates a number of properties that have to be satisfied by a utility maximization problem. These fundamental properties ensure the existence of an optimal parameter setup. Moreover, they establish a monotonicity of the behavior of the upper layers with respect to the physical performance, as discussed in Sect. 2.1.

While existence of an optimal parameter setup is ensured by Definition 2.2.1, it is by no means guaranteed that an efficient algorithm for finding such an optimal parameter setup exists. In contrast, throughout this chapter, it will become clear that additional properties are needed. From the viewpoint of optimization theory, the optimal parameter setup represents a global maximizer of the maximization problem (2.30). If such a global maximizer is to be found for large problem dimension,[1] problem convexity represents an indispensable structural requirement [20]. In the simplest case, both utility model and physical layer model are convex. Under a few additional assumptions, the utility maximization problem can then be solved in polynomial time using standard methods. This special case is discussed in Sect. 3.1.

Advanced physical layer techniques and application-oriented utility functions usually result in nonconvex models. Accordingly, in this monograph, cases where at least one of the models is nonconvex are investigated. In case that the utility model is convex, it is possible to obtain a convex optimization problem by changing the optimization domain. Section 3.3 shows how the reformulated problem can be

[1]Recall from Definition 2.2.1 that a parameter setup $\boldsymbol{x}_1$ is an element of the parameter space $\mathcal{D}_1$. Intuitively, the problem dimension is then the dimension of $\mathcal{D}_1$. However, it may not be possible to assign a dimension to an arbitrary metric space $\mathcal{D}_1$. In the following, it is assumed that the set $\mathcal{D}_1$ is a subset (but not necessarily a subspace) of a finite-dimensional vector space $\mathcal{V}$, which then immediately defines the problem dimension as the dimension of $\mathcal{V}$.

J. Brehmer, *Utility Maximization in Nonconvex Wireless Systems*, Foundations in Signal Processing, Communications and Networking 5, DOI 10.1007/978-3-642-17438-4_3,

solved based on Lagrange duality. In addition to requiring convexity of the utility layer, this approach relies on two crucial assumptions: the physical layer provides a scheduling mode (cf. Sect. 2.3.5) and there exists a method to maximize a weighted sum of rates.

If the utility model is nonconvex or if scheduling is not feasible, it is generally not possible to obtain a convex problem formulation. By Definition 2.2.1, a utility maximization problem offers a monotonic structure at the utility level, i.e., both u and h are increasing functions. The monotonic structure can be extracted by again changing the optimization domain. Based on this reformulation, methods for deterministic global optimization can be employed to find a globally optimal parameter setup, as discussed in Sect. 3.4. This approach makes no additional assumptions concerning the properties of a utility maximization, but relies on a method to compute the intersection between a line and the boundary of the rate region. By the reformulation, the dimension of the optimization problem is reduced from the dimension of the physical layer parameter space to the number of links in the network. Due to their complexity, however, global optimization methods are practically feasible only for relatively small problem dimension, i.e., for networks of small size.

For larger problem dimension and in the absence of convexity, it is practically not possible to find a parameter setup that is guaranteed to be globally optimal. As a consequence, the notion of a solution of a utility maximization problem has to be loosened by accepting parameter setups as a solution that fulfill a necessary optimality condition which is based on local information. Methods that find such local solutions are discussed in Sect. 3.5.

3.1 A Simple Utility Maximization Problem

Let $(u, h, \boldsymbol{r}, \mathcal{X}_1)$ be a utility maximization problem according to Definition 2.2.1. Assume that u, h, and $\boldsymbol{r}$ are concave functions, and that

$$\mathcal{X}_1 = \left\{\boldsymbol{x}_1 \in \mathbb{R}^N : \boldsymbol{c}(\boldsymbol{x}) \geq \boldsymbol{0}\right\}, \tag{3.1}$$

where $\boldsymbol{c} : \mathbb{R}^N \to \mathbb{R}^J$ is a concave function. The definition of $\mathcal{X}_1$ in (3.1) implies that $\mathcal{D}_1 \subseteq \mathbb{R}^N$, i.e., the physical layer has N real-valued parameters.

By definition of a utility maximization problem, u and h are increasing, thus, with Proposition A.5.2, the functions $u \circ \boldsymbol{r}$ and $h \circ \boldsymbol{r}$ are concave. Accordingly, the utility maximization problem can be written as

$$\max_{\boldsymbol{x}_1 \in \mathbb{R}^N} (u \circ \boldsymbol{r})(\boldsymbol{x}_1) \quad \text{s.t.} \quad (h \circ \boldsymbol{r})(\boldsymbol{x}_1) \geq 0, \boldsymbol{c}(\boldsymbol{x}_1) \geq \boldsymbol{0}. \tag{3.2}$$

Problem (3.2) corresponds to a convex problem in standard form: A concave cost function $u \circ \boldsymbol{r}$ is maximized over a convex feasible set

$$\mathcal{F}_1 = \{\boldsymbol{x}_1 \in \mathcal{D}_1 : (h \circ \boldsymbol{r}, \boldsymbol{c})(\boldsymbol{x}_1) \geq \boldsymbol{0}\}, \tag{3.3}$$

with a concave constraint function $(h \circ \boldsymbol{r}, \boldsymbol{c})$. By Definition 2.2.1, the feasible set $\mathcal{F}_1$ is nonempty.

If it is additionally assumed that:

(1) There exists a means to compute a subgradient of c_j, $j = 1, \dots, J$, at any $\boldsymbol{x}_1 \in \mathbb{R}^N$ (which implies the existence of a subgradient of c_j at $\boldsymbol{x}_1$, $\forall \boldsymbol{x}_1 \in \mathbb{R}^N$).
(2) There exists a means to compute a subgradient of $h \circ \boldsymbol{r}$ at any $\boldsymbol{x}_1 \in \mathcal{D}_1$.
(3) There exists a means to compute a subgradient of $u \circ \boldsymbol{r}$ at any $\boldsymbol{x}_1 \in \mathcal{F}_1$.
(4) And the set $\mathcal{F}_1$, besides being compact and convex, has nonempty interior.

Then Problem (3.2) can be solved in polynomial time by a universal algorithm, the Ellipsoid method [23]. Note that the *separation oracle* [23] required by the Ellipsoid method can be constructed from the subgradients provided by Assumptions (1) and (2). If additional assumptions are made, one may also apply more efficient Interior Point methods to solve (3.2) [23]. In other words, solving a utility maximization problem that offers the convexity structure required for the convex standard form (3.2) is in general straightforward: The optimal parameter setup can be found by feeding the problem data into a general-purpose solver. Evidently, the last statement only holds if a centralized algorithm can be used to solve the problem. Even for convex problems, the development of distributed algorithms is generally a challenging task [17, 18].

From the perspective of advanced physical layer techniques, however, it is rather unlikely that a utility maximization problem offers the required convexity structure: None of the physical layer models discussed in Chap. 4 has a concave rate function $\boldsymbol{r}$. Thus, even if the utility model (u, h) is convex, more elaborate solution strategies are required. Quite obviously, the same holds if the utility model is nonconvex.

3.2 Rate Space Formulation

Reconsider Problem (3.2) for the case that the rate function $\boldsymbol{r}$ is not concave. In general, neither the function $u \circ \boldsymbol{r}$ nor the function $h \circ \boldsymbol{r}$ is concave, even if u and h are assumed to be concave. Moreover, except for the special case that $\boldsymbol{r}$ is increasing, neither the function $u \circ \boldsymbol{r}$ nor the function $h \circ \boldsymbol{r}$ is increasing. Thus, in general, Problem (3.2) does not offer any additional structure with respect to the optimization variable $\boldsymbol{x}_1$.

A problem that is closely related to the problem of finding an optimal parameter setup is the rate space problem, which was introduced in Sect. 2.3. In the rate space formulation, the optimization domain is changed by using rates instead of parameters as the optimization variables:

$$\max_{\boldsymbol{s} \in \mathcal{R}} u(\boldsymbol{s}) \quad \text{s.t.} \quad h(\boldsymbol{s}) \geq \boldsymbol{0}, \tag{3.4}$$

with rate region $\mathcal{R} = \boldsymbol{r}(\mathcal{X}_1)$. According to Definition 2.2.1, the functions u and h are increasing, while $\mathcal{R}$ is a compact set. Accordingly, the rate space problem is a monotonic optimization problem [24]. In other words, the rate space problem offers a rich structure with respect to the optimization variable $\boldsymbol{s}$. This property is exploited in Sect. 3.4.

Moreover, as discussed in Sect. 2.3.5, a convex rate region can be obtained by simply taking the convex hull of a nonconvex rate region $\mathcal{R}$. Thus, if the functions u and h are concave, the rate space problem can always be turned into a convex optimization problem by adding a time-sharing mode at the physical layer. Still, it is important to note that even if the rate space problem is convex, it cannot be solved with Interior Point methods, due to the fact that the constraint on $\boldsymbol{s}$ is in terms of membership in the set $\mathcal{R}$, and not in terms of differentiable constraint functions. On the contrary, there are cases where time-sharing makes finding an optimal parameter setup more difficult. In the following, it is shown that in the case of a convex rate space problem, tackling the rate space problem instead of the original utility maximization problem (2.30) is beneficial if solving a weighted sum rate maximization problem is significantly simpler than solving the utility maximization problem directly. In this case, the key to solving a convex rate space problem is Lagrange duality, as shown in Sect. 3.3.

3.3 Lagrange Duality

Let $(u, h, \boldsymbol{r}, \mathcal{X}_1)$ denote a utility maximization problem. It is assumed that the rate region

$$\mathcal{C} = \boldsymbol{r}(\mathcal{X}_1)$$

is convex and proper. The symbol $\mathcal{C}$ instead of $\mathcal{R}$ is used for the rate region to emphasize the fact that the rate region is convex. Moreover, it is assumed that $u : \mathbb{R}_+^K \to \mathbb{R}$ is concave and h is the point-wise minimum over the component functions of a concave increasing function $\boldsymbol{h} : \mathbb{R}_+^K \to \mathbb{R}^M$, for some $M \geq 1$, i.e.,

$$h = \min_m h_m(\cdot). \tag{3.5}$$

The function h is concave and increasing. By Proposition A.5.1, h is continuous, thus a function h defined by (3.5) satisfies the requirements of a utility maximization problem. The assumption that h is the point-wise minimum over the component functions of $\boldsymbol{h}$ is motivated as follows: Taking the point-wise minimum over the component functions leads to a loss of information. As a result, depending on the algorithm used to solve the rate space problem, using $\boldsymbol{h}$ instead of h may result in a more efficient solution. Accordingly, throughout this section, the function $\boldsymbol{h}$ is used in the development of a solution method. Clearly, the case $M = 1$ is included, thus using $\boldsymbol{h}$ instead of h does not represent a limitation.

The rate region is proper, hence the utopia point $\boldsymbol{s}^{\mathrm{U}}$ of $\mathcal{C}$ is well-defined. Define

$$\mathcal{Z} = \left[\mathbf{0}, \boldsymbol{s}^{\mathrm{U}}\right]. \tag{3.6}$$

According to Corollary 2.3.9, $\mathcal{C} \subseteq \mathcal{Z}$. Moreover, let

$$\mathcal{H} = \{\boldsymbol{z} \in \mathcal{Z} : \boldsymbol{h}(\boldsymbol{z}) \geq \mathbf{0}\}. \tag{3.7}$$

By monotonicity of $\boldsymbol{h}$,

$$\boldsymbol{h}(\mathbf{0}) \geq \mathbf{0} \implies \mathcal{H} = \mathcal{Z} \supseteq \mathcal{C}.$$

In other words, if $\boldsymbol{h}(\mathbf{0}) \geq \mathbf{0}$, then the constraint $\boldsymbol{h}(\boldsymbol{s}) \geq \mathbf{0}$ is redundant. Accordingly, two cases are considered, $\boldsymbol{h}(\mathbf{0}) \geq \mathbf{0}$ and $\boldsymbol{h}(\mathbf{0}) \notin \mathbb{R}_+^K$.

Using the function $\boldsymbol{h}$ instead of the point-wise minimum h, the rate space problem is given by

$$\max_{\boldsymbol{s}} u(\boldsymbol{s}) \quad \text{s.t.} \quad \boldsymbol{h}(\boldsymbol{s}) \geq \mathbf{0}, \boldsymbol{s} \in \mathcal{C}. \tag{3.8}$$

A generic solution method to Problem (3.8) is presented in Sect. 3.3.1, which does not make any further assumptions concerning u and $\boldsymbol{h}$. In the special case that both u and $\boldsymbol{h}$ are affine, a more efficient solution is possible, which is discussed in Sect. 3.3.4.

3.3.1 Generic Approach

Consider the following problem:

$$\max_{(\boldsymbol{s},\boldsymbol{z})} u(\boldsymbol{z}) \quad \text{s.t.} \quad \boldsymbol{h}(\boldsymbol{z}) \geq \mathbf{0}, \boldsymbol{s} - \boldsymbol{z} \geq \mathbf{0}, (\boldsymbol{s}, \boldsymbol{z}) \in \mathcal{C} \times \mathcal{Z}. \tag{3.9}$$

The set $\mathcal{C} \times \mathcal{Z}$ is compact, and both $\mathbb{R}_+^K \times \boldsymbol{h}^{-1}(\mathbb{R}_+^K)$ and $\{(\boldsymbol{s}, \boldsymbol{z}) \in \mathbb{R}_+^K \times \mathbb{R}_+^K : \boldsymbol{s} - \boldsymbol{z} \geq \mathbf{0}\}$ are closed sets. Moreover, u is continuous. Hence, Problem (3.9) is well-defined and a maximizer exists.

Proposition 3.3.1. *Let* $(\boldsymbol{s}^*, \boldsymbol{z}^*)$ *denote a maximizer of Problem* (3.9). *Then* $\boldsymbol{s}^*$ *is a maximizer of Problem* (3.8), *and* $u(\boldsymbol{z}^*) = u(\boldsymbol{s}^*)$.

Proof. By feasibility of $(\boldsymbol{s}^*, \boldsymbol{z}^*)$, $\boldsymbol{s}^* \geq \boldsymbol{z}^*$. By monotonicity of u, $u(\boldsymbol{z}^*) \leq u(\boldsymbol{s}^*)$. The rate region is proper, thus $\mathcal{C} \subseteq \mathcal{Z}$ and $\boldsymbol{s}^* \in \mathcal{Z}$. From the monotonicity of $\boldsymbol{h}$, it follows that $(\boldsymbol{s}^*, \boldsymbol{s}^*)$ is also feasible, thus $u(\boldsymbol{z}^*) = u(\boldsymbol{s}^*)$. From the fact that the maximum is attained by a point with $\boldsymbol{s} = \boldsymbol{z}$, it follows that $\boldsymbol{s}^*$ is a maximizer of Problem (3.8). □

According to Proposition 3.3.1, Problem (3.8) can be solved by solving Problem (3.9). In Problem (3.9), upper layers and physical layer are only coupled via the constraint $\boldsymbol{s} - \boldsymbol{z} \geq \boldsymbol{0}$. The benefit of solving Problem (3.9) lies in the fact that this linear coupling leads to a decoupling effect in the dual domain and a reduction of the physical layer subproblem to solving a weighted sum of rates, as discussed next.

Problem (3.9) is solved by Lagrange duality. Only the constraint $\boldsymbol{s} - \boldsymbol{z} \geq \boldsymbol{0}$ is dualized. In view of Proposition 3.3.1, it is of primary interest to compute a maximizer $(\boldsymbol{s}^*, \boldsymbol{z}^*)$ of Problem (3.9). Such a maximizer can be obtained from the dual problem by means of primal recovery, under the condition that *strong duality* holds [25]. Here, only the case where strong duality is established by means of Slater's condition [25] is considered. Recalling that only the constraint $\boldsymbol{s} - \boldsymbol{z} \geq \boldsymbol{0}$ is dualized, for Slater's condition to hold there has to exist a feasible $(\boldsymbol{s}, \boldsymbol{z})$ such that

$$\boldsymbol{s} - \boldsymbol{z} \gg \boldsymbol{0}. \tag{3.10}$$

If $\boldsymbol{h}(\boldsymbol{0}) \geq \boldsymbol{0}$, then the assumption that $\mathcal{C}$ is proper is sufficient for strong duality to hold: If $\mathcal{C}$ is proper, there exists $\boldsymbol{s}' \in \mathcal{C}$ such that $\boldsymbol{s}' \gg \boldsymbol{0}$. Moreover, $\boldsymbol{0} \in \mathcal{H}$, thus $(\boldsymbol{s}', \boldsymbol{0})$ is feasible and fulfills Condition (3.10).

In the second case where $\boldsymbol{h}(\boldsymbol{0}) \notin \mathbb{R}_+^K$, it depends on the problem instance whether Slater's condition holds. Note that a feasible $(\boldsymbol{s}, \boldsymbol{z})$ that satisfies Condition (3.10) exists if

$$\operatorname{int}(\mathcal{C} \cap \mathcal{H}) \neq \emptyset. \tag{3.11}$$

By definition, a utility maximization problem is feasible. Feasibility, however, only implies

$$\mathcal{C} \cap \mathcal{H} \neq \emptyset,$$

which clearly is a weaker condition than Condition (3.11). An algorithm for testing whether Slater's condition holds is provided in Sect. 3.3.3.

In the following, it is assumed that strong duality holds for Problem (3.9). By dualizing the constraint $\boldsymbol{s} - \boldsymbol{z} \geq \boldsymbol{0}$, the Lagrangian is given by

$$L(\boldsymbol{s}, \boldsymbol{z}, \boldsymbol{\lambda}) = u(\boldsymbol{z}) - \boldsymbol{\lambda}^{\mathrm{T}} \boldsymbol{z} + \boldsymbol{\lambda}^{\mathrm{T}} \boldsymbol{s}. \tag{3.12}$$

The dual function $q : \mathbb{R}_+^K \to \mathbb{R}$ follows as

$$q(\boldsymbol{\lambda}) = \max_{(\boldsymbol{s},\boldsymbol{z}) \in \mathcal{C} \times \mathcal{H}} L(\boldsymbol{s}, \boldsymbol{z}, \boldsymbol{\lambda}) = \max_{\boldsymbol{z} \in \mathcal{H}} u(\boldsymbol{z}) - \boldsymbol{\lambda}^{\mathrm{T}} \boldsymbol{z} + \max_{\boldsymbol{s} \in \mathcal{C}} \boldsymbol{\lambda}^{\mathrm{T}} \boldsymbol{s}. \tag{3.13}$$

According to (3.13), the evaluation of the dual function at $\boldsymbol{\lambda} \geq \mathbf{0}$ decouples into two subproblems. It is assumed that the first subproblem

$$\max_{\boldsymbol{z} \in \mathcal{Z}} u(\boldsymbol{z}) - \boldsymbol{\lambda}^{\mathrm{T}} \boldsymbol{z} \quad \text{s.t.} \quad \boldsymbol{h}(\boldsymbol{z}) \geq \mathbf{0} \tag{3.14}$$

can be solved by standard methods for convex optimization (note that the constraint $\boldsymbol{z} \in \mathcal{Z}$ corresponds to a simple box constraint). The second subproblem

$$\max_{\boldsymbol{s} \in \mathcal{C}} \boldsymbol{\lambda}^{\mathrm{T}} \boldsymbol{s} \tag{3.15}$$

is a weighted sum rate maximization problem. As a result, if the physical layer provides a means to solve Problem (3.15) for a given $\boldsymbol{\lambda} \in \mathbb{R}_{+}^{K}$, then the dual function can be evaluated.

Lagrange duality can be considered a standard tool in the optimization of wireless systems. The decomposition of the dual function observed in (3.13) is an instance of a *dual decomposition*. One of the many fields where dual decomposition methods have been used to exploit a separable structure is network utility maximization [14, 26, 27]. Some further examples of Lagrange duality-based approaches to wireless system optimization are discussed in Sect. 3.3.6.

3.3.2 Solving the Dual Problem

In this work, it is assumed that the use of centralized optimization algorithms is permissible. As a result, the dual problem can be solved by standard methods, provided that a method for solving the weighted sum rate maximization problem (3.15) is available. In this case, a basic issue in the choice of a solution method is the question whether the dual function is differentiable. In the following, it is shown that the dual function is nondifferentiable if the convex rate region $\mathcal{C}$ is obtained by taking the convex hull of a nonconvex rate region.

Define the set of maximizers of Problems (3.14) and (3.15) as follows:

$$\bar{\mathcal{Z}}(\boldsymbol{\lambda}) = \operatorname*{argmax}_{\boldsymbol{z} \in \mathcal{Z}} u(\boldsymbol{z}) - \boldsymbol{\lambda}^{\mathrm{T}} \boldsymbol{z} \quad \text{s.t.} \quad \boldsymbol{h}(\boldsymbol{z}) \geq \mathbf{0}, \tag{3.16}$$

$$\bar{\mathcal{C}}(\boldsymbol{\lambda}) = \operatorname*{argmax}_{\boldsymbol{s} \in \mathcal{C}} \boldsymbol{\lambda}^{\mathrm{T}} \boldsymbol{s}. \tag{3.17}$$

For every pair $(\boldsymbol{s}, \boldsymbol{z}) \in \bar{\mathcal{C}}(\boldsymbol{\lambda}) \times \bar{\mathcal{Z}}(\boldsymbol{\lambda})$, $\boldsymbol{d} = \boldsymbol{s} - \boldsymbol{z}$ is a subgradient of q at $\boldsymbol{\lambda}$ [22]. Accordingly,

$$\partial q(\boldsymbol{\lambda}) \supseteq \left\{\boldsymbol{s} - \boldsymbol{z} : (\boldsymbol{s}, \boldsymbol{z}) \in \bar{\mathcal{C}}(\boldsymbol{\lambda}) \times \bar{\mathcal{Z}}(\boldsymbol{\lambda})\right\}, \tag{3.18}$$

where $\partial q(\boldsymbol{\lambda})$ is the set of subgradients of q at $\boldsymbol{\lambda}$. The dual function is differentiable at $\boldsymbol{\lambda}$ if and only if $\partial q(\boldsymbol{\lambda})$ is a singleton [22]. In general, though, neither $\bar{\mathcal{Z}}(\boldsymbol{\lambda})$ nor $\bar{\mathcal{C}}(\boldsymbol{\lambda})$ can be assumed to be singletons. In particular, if $\mathcal{C}$ is obtained from a nonconvex proper rate region via time sharing, then Proposition 3.3.2 shows that there exists $\boldsymbol{\lambda} \in \mathbb{R}_+^K$ such that $\bar{\mathcal{C}}(\boldsymbol{\lambda})$ is not a singleton.

Proposition 3.3.2. *Let $\mathcal{R}$ be a nonconvex proper rate region. If $\mathcal{C} = \operatorname{conv}(\mathcal{R})$, then there exists $\boldsymbol{\lambda} \in \mathbb{R}_+^K$ such that $\bar{\mathcal{C}}(\boldsymbol{\lambda})$ is not a singleton.*

Proof. By Proposition A.5.10, $\operatorname{wpar}(\mathcal{C}) \setminus (\operatorname{wpar}(\mathcal{C}) \cap \mathcal{R}) \neq \emptyset$. Let $\boldsymbol{s} \in \operatorname{wpar}(\mathcal{C}) \setminus (\operatorname{wpar}(\mathcal{C}) \cap \mathcal{R})$. As $\boldsymbol{s} \in \operatorname{conv}(\mathcal{R})$, $\boldsymbol{s}$ can be written as a convex combination of elements of $\mathcal{R}$:

$$\boldsymbol{s} = \sum_{w=1}^{W} \alpha_w \boldsymbol{s}^{(w)}, \tag{3.19}$$

with $\boldsymbol{s}^{(w)} \in \mathcal{R}$, $\alpha_w > 0$, and $\sum_{w=1}^{W} \alpha_w = 1$. From $\boldsymbol{s} \notin \mathcal{R}$ it follows that $W > 1$ and

$$\boldsymbol{s} \neq \boldsymbol{s}^{(w)}, \forall w = 1, \ldots, W. \tag{3.20}$$

As $\boldsymbol{s} \in \operatorname{wpar}(\mathcal{C})$, by Proposition A.5.6, there exists $\boldsymbol{\lambda} \in \mathbb{R}_+^K$ such that

$$\boldsymbol{\lambda}^{\mathrm{T}} \boldsymbol{s} = \max_{\boldsymbol{y} \in \mathcal{C}} \boldsymbol{\lambda}^{\mathrm{T}} \boldsymbol{y}. \tag{3.21}$$

In other words, $\boldsymbol{s} \in \bar{\mathcal{C}}(\boldsymbol{\lambda})$. By Proposition A.5.8,

$$\boldsymbol{\lambda}^{\mathrm{T}} \boldsymbol{s} = \boldsymbol{\lambda}^{\mathrm{T}} \boldsymbol{s}^{(w)}, \forall w = 1, \ldots, W.$$

From $\boldsymbol{s}^{(w)} \in \mathcal{C}$ it follows that

$$\bar{\mathcal{C}}(\boldsymbol{\lambda}) \supseteq \left\{\boldsymbol{s}, \boldsymbol{s}^{(1)}, \ldots, \boldsymbol{s}^{(W)}\right\}.$$

Equation (3.20) implies that $\bar{\mathcal{C}}(\boldsymbol{\lambda})$ is not a singleton. □

The dual function q is always convex, but, as evidenced by (3.18) and Proposition 3.3.2, not necessarily differentiable. Accordingly, methods for nondifferentiable convex optimization are applied to solve the dual problem

$$\min_{\boldsymbol{\lambda} \geq \mathbf{0}} q(\boldsymbol{\lambda}). \tag{3.22}$$

The set of methods available for solving the dual problem is manifold, including subgradient methods [25, 28, 29], cutting-plane methods [25, 29, 30], the Ellipsoid method [23], and bundle methods [29]. Note that in Problem (3.22), the feasible set is unbounded. The Ellipsoid method and most cutting-plane methods require a problem formulation

$$\min_{\boldsymbol{\lambda}\in\Lambda} q(\boldsymbol{\lambda}), \tag{3.23}$$

where $\Lambda \subset \mathbb{R}_+^K$ is compact and contains a minimizer of the dual problem (3.22). By (3.13),

$$q(\boldsymbol{\lambda}) \geq L(\boldsymbol{s}, \boldsymbol{z}, \boldsymbol{\lambda}), \forall(\boldsymbol{s}, \boldsymbol{z}) \in \mathcal{C} \times \mathcal{H}. \tag{3.24}$$

A set Λ with the desired properties can be derived from a pair $(\boldsymbol{s}', \boldsymbol{z}') \in \mathcal{C} \times \mathcal{H}$ with $\boldsymbol{s}' - \boldsymbol{z}' \gg \mathbf{0}$. Obviously, the problem of finding such a pair is closely related to verifying Slater's condition. Accordingly, depending on the value $\boldsymbol{h}(\mathbf{0})$, the proofs of Propositions 2.3.11 and 3.3.3 provide algorithms for computing such a pair. Moreover, let $\boldsymbol{\lambda}^*$ denote a minimizer of Problem (3.22). According to (3.24) and the optimality of $\boldsymbol{\lambda}^*$,

$$L(\boldsymbol{s}', \boldsymbol{z}', \boldsymbol{\lambda}^*) = u(\boldsymbol{z}') + (\boldsymbol{\lambda}^*)^{\mathrm{T}}(\boldsymbol{s}' - \boldsymbol{z}') \leq q(\boldsymbol{\lambda}^*) \leq q(\mathbf{1}). \tag{3.25}$$

Rearranging (3.25) yields

$$(\boldsymbol{\lambda}^*)^{\mathrm{T}}(\boldsymbol{s}' - \boldsymbol{z}') = \sum_{k=1}^{K} \lambda_k^*(s_k' - z_k') \leq q(\mathbf{1}) - u(\boldsymbol{z}').$$

Moreover, $\lambda_k^*(s_k' - z_k') \geq 0, \forall k$, hence

$$\lambda_k^*(s_k' - z_k') \leq q(\mathbf{1}) - u(\boldsymbol{z}').$$

Observing that $\boldsymbol{s}' - \boldsymbol{z}' \gg \mathbf{0}$ yields an upper bound on λ_k^*:

$$\lambda_k^* \leq \frac{q(\mathbf{1}) - u(\boldsymbol{z}')}{s_k' - z_k'}. \tag{3.26}$$

Accordingly, a simple box constraint on $\boldsymbol{\lambda}$ is given by

$$\Lambda = \left\{\boldsymbol{\lambda} \in \mathbb{R}^K : 0 \leq \lambda_k \leq \frac{q(\mathbf{1}) - u(\boldsymbol{z}')}{s_k' - z_k'}\right\}. \tag{3.27}$$

All of the methods for nondifferentiable convex optimization mentioned above have in common that they generate a sequence of iterates $\boldsymbol{\lambda}^{(n)}$ that converges to a minimizer $\boldsymbol{\lambda}^*$ of the dual problem (3.22). Moreover, they can be represented by a black-box model [22], taking as input a pair $(q(\boldsymbol{\lambda}^{(n)}), \boldsymbol{d}^{(n)})$, where $\boldsymbol{d}^{(n)}$ is *any* subgradient of g at $\boldsymbol{\lambda}^{(n)}$, i.e.,

$$\boldsymbol{d}^{(n)} \in \partial q(\boldsymbol{\lambda}^{(n)}), \tag{3.28}$$

and returning as output the next iterate $\boldsymbol{\lambda}^{(n+1)}$.

Assume that $\mathcal{C}$ is obtained from a nonconvex proper rate region $\mathcal{R}$ by time-sharing. By Corollary A.5.9

$$\max_{\boldsymbol{s}\in\mathcal{C}} \boldsymbol{\lambda}^{\mathrm{T}}\boldsymbol{s} = \max_{\boldsymbol{s}\in\mathcal{R}} \boldsymbol{\lambda}^{\mathrm{T}}\boldsymbol{s}. \tag{3.29}$$

Accordingly, time-sharing is not needed to compute $q(\boldsymbol{\lambda})$. Define

$$\bar{\mathcal{R}}(\boldsymbol{\lambda}) = \operatorname*{argmax}_{\boldsymbol{s}\in\mathcal{R}} \boldsymbol{\lambda}^{\mathrm{T}}\boldsymbol{s}. \tag{3.30}$$

By (3.29) and $\mathcal{R} \subseteq \mathcal{C}$,

$$\bar{\mathcal{R}}(\boldsymbol{\lambda}) \subseteq \bar{\mathcal{C}}(\boldsymbol{\lambda}). \tag{3.31}$$

As a result, any $\boldsymbol{d} \in \bar{\mathcal{R}}(\boldsymbol{\lambda}) \times \bar{\mathcal{Z}}(\boldsymbol{\lambda})$ is a subgradient of q at $\boldsymbol{\lambda}$. Thus, time-sharing is not needed to obtain a pair $(q(\boldsymbol{\lambda}^{(n)}), \boldsymbol{d}^{(n)})$. In other words, the dual solution $\boldsymbol{\lambda}^*$ can be computed without time-sharing.

Under the assumption that strong duality holds, primal solution can be obtained by standard methods for primal recovery [25]. Similar to the method for computing the dual solution, primal recovery can be modeled by a black box model. The input at each iteration is a quadrupel $((\bar{\boldsymbol{s}}^{(n)}, \bar{\boldsymbol{z}}^{(n)}), q(\boldsymbol{\lambda}^{(n)}), \boldsymbol{d}^{(n)}, \boldsymbol{\lambda}^{(n)})$, where

$$(\bar{\boldsymbol{s}}^{(n)}, \bar{\boldsymbol{z}}^{(n)}) \in \bar{\mathcal{C}}(\boldsymbol{\lambda}^{(n)}) \times \bar{\mathcal{Z}}(\boldsymbol{\lambda}^{(n)}). \tag{3.32}$$

The output is a feasible solution $(\boldsymbol{s}^{(n)}, \boldsymbol{z}^{(n)})$. The sequence of outputs $(\boldsymbol{s}^{(n)}, \boldsymbol{z}^{(n)})$ converges to the primal solution $(\boldsymbol{s}^*, \boldsymbol{z}^*)$.

In case of $\mathcal{C}$ being obtained from a nonconvex rate region by time-sharing, at each iteration, it is possible to choose

$$\bar{\boldsymbol{s}}^{(n)} \in \bar{\mathcal{R}}(\boldsymbol{\lambda}^{(n)}),$$

an immediate consequence of (3.31) and (3.32). As a result, time-sharing is also not needed for computing the input value $\bar{\boldsymbol{s}}^{(n)}$ for the primal recovery method. In contrast, the optimal solution $(\boldsymbol{s}^*, \boldsymbol{z}^*)$ may require time-sharing, i.e., there exist u, h, and $\mathcal{R}$ such that

$$\boldsymbol{s}^* \in \mathcal{C} \setminus \mathcal{R}.$$

Evidently, primal recovery implicitly performs time-sharing, if needed.

3.3.3 Verifying Slater's Condition

As discussed in Sect. 3.3.1, strong duality holds if $\boldsymbol{h}(\mathbf{0}) \geq \mathbf{0}$, which is equivalent to a problem without constraint $\boldsymbol{h}(\boldsymbol{s}) \geq \mathbf{0}$. In the second case where $\boldsymbol{h}(\mathbf{0}) \notin \mathbb{R}_+^K$, one possible way to test whether Condition (3.11) holds is to solve the problem

$$\max_{(\boldsymbol{s},\boldsymbol{z})} h(\boldsymbol{z}) \quad \text{s.t.} \quad \boldsymbol{s} - \boldsymbol{z} \geq \mathbf{0}, (\boldsymbol{s}, \boldsymbol{z}) \in \mathcal{C} \times \mathcal{Z}, \tag{3.33}$$

with h as defined in (3.5). Note that Problem (3.33) is the same problem type as Problem (3.9), due to the fact that the function h is concave and increasing. As discussed in Sect. 3.3.1, Slater's condition holds for Problem (3.33), due to the assumption that $\mathcal{C}$ is convex and proper. Accordingly, Problem (3.33) can be solved by Lagrange duality, as described in Sect. 3.3.1.

Proposition 3.3.3. *Let* $(\check{\boldsymbol{s}}, \check{\boldsymbol{z}})$ *be a maximizer of Problem* (3.33)*. Slater's condition holds for Problem* (3.9) *if*

$$h(\check{\boldsymbol{z}}) > 0. \tag{3.34}$$

In addition, if $\boldsymbol{h}$ *is strongly increasing, Condition* (3.34) *is necessary for Slater's condition to hold.*

Proof. To show that Condition (3.34) is a sufficient condition, a pair $(\boldsymbol{s}', \boldsymbol{z}') \in \mathcal{C} \times \mathcal{H}$ satisfying $\boldsymbol{z}' \ll \boldsymbol{s}'$ is constructed.

Due to the fact that $\boldsymbol{h}$ is increasing, $(\check{\boldsymbol{s}}, \check{\boldsymbol{s}})$ is also a maximizer, and $h(\check{\boldsymbol{s}}) = h(\check{\boldsymbol{z}})$. Moreover, $\boldsymbol{h}(\mathbf{0}) \notin \mathbb{R}_+^K$ is equivalent to

$$h(\mathbf{0}) = \min_m h_m(\mathbf{0}) < 0.$$

Let $h(\check{\boldsymbol{s}}) = h(\check{\boldsymbol{z}}) > \mathbf{0}$. It follows that $\check{\boldsymbol{s}} \neq \mathbf{0}$. From the continuity of h and $\check{\boldsymbol{s}} \neq \mathbf{0}$ it follows that there exists $t \in (0, 1)$ such that

$$h(t\check{\boldsymbol{s}}) = 0. \tag{3.35}$$

Clearly, $t\check{\boldsymbol{s}} \in \mathcal{H}$. Moreover, $t\check{\boldsymbol{s}} \leq \check{\boldsymbol{s}}$. Still, if any of the entries of $\check{\boldsymbol{s}}$ is equal to zero, the relation $t\check{\boldsymbol{s}} \ll \check{\boldsymbol{s}}$ does not hold. Accordingly, in the case that $\check{\boldsymbol{s}}$ has zero entries, a vector $\boldsymbol{s}' \in \mathcal{C}$ is constructed such that $t\check{\boldsymbol{s}} \ll \boldsymbol{s}'$. Define an index set $\mathcal{I}$ as follows:

$$\mathcal{I} = \{k : \check{s}_k = 0\}.$$

Let $\{\boldsymbol{s}^1, \ldots, \boldsymbol{s}^K\}$ denote the set of single-link points of $\mathcal{C}$. Define $\boldsymbol{s}'$ as follows:

$$\boldsymbol{s}' = \begin{cases} \check{\boldsymbol{s}}, & \mathcal{I} = \emptyset, \\ \frac{1+t}{2}\check{\boldsymbol{s}} + \sum\limits_{k \in \mathcal{I}} \frac{1-t}{2\,\mathrm{card}(\mathcal{I})} \boldsymbol{s}^k, & \mathcal{I} \neq \emptyset. \end{cases} \tag{3.36}$$

By convexity of $\mathcal{C}$ and $0 < t < 1$, $\boldsymbol{s}' \in \mathcal{C}$. Moreover, $t\check{\boldsymbol{s}} \ll \boldsymbol{s}'$: If $\mathcal{I} = \emptyset$, the result is obvious. Assume $\mathcal{I} \neq \emptyset$. First, let $k \notin \mathcal{I}$. Then $\check{s}_k > 0$ and $t < \frac{1+t}{2}$, thus $t\check{s}_k < s'_k$. Second, let $k \in \mathcal{I}$. Then $\check{s}_k = 0$ and $s^k_k > 0$, hence again $t\check{s}_k < s'_k$. As a result, $(\boldsymbol{s}', \boldsymbol{z}') = (\boldsymbol{s}', t\check{\boldsymbol{s}}) \in \mathcal{C} \times \mathcal{H}$ fulfills Slater's condition.

To show that Condition (3.34) is a necessary condition if $\boldsymbol{h}$ is strongly increasing, assume that $(\boldsymbol{s}', \boldsymbol{z}') \in \mathcal{C} \times \mathcal{H}$, $\boldsymbol{z}' \ll \boldsymbol{s}'$, and $h(\check{\boldsymbol{z}}) \leq 0$. Note that

$$h(\check{\boldsymbol{z}}) = \max\{h(\boldsymbol{s}) : \boldsymbol{s} \in \mathcal{C}\}.$$

If $\boldsymbol{h}$ is strongly increasing,

$$\boldsymbol{z}' \ll \boldsymbol{s}' \implies \boldsymbol{h}(\boldsymbol{z}') \ll \boldsymbol{h}(\boldsymbol{s}').$$

But

$$\begin{aligned} \boldsymbol{h}(\boldsymbol{z}') \ll \boldsymbol{h}(\boldsymbol{s}') &\implies h(\boldsymbol{z}') < h(\boldsymbol{s}') \\ &\implies h(\boldsymbol{z}') < 0 \\ &\implies \boldsymbol{z}' \notin \mathcal{H}, \end{aligned}$$

a contradiction. □

Note that Problem (3.33) can also be used as a feasibility test for Problem (3.4): if $h(\check{\boldsymbol{z}}) < 0$, then Problem (3.4) is infeasible, otherwise it is feasible.

3.3.4 Special Case: Affine *u* and *h*

In the special case that both u and $\boldsymbol{h}$ are affine, a more efficient solution approach than the generic method presented in Sect. 3.3.1 is possible.

Affinity of u and $\boldsymbol{h}$ implies that u and $\boldsymbol{h}$ can be written as

$$u(\boldsymbol{s}) = \boldsymbol{a}^{\mathrm{T}}\boldsymbol{s} + b, \tag{3.37}$$

$$\boldsymbol{h}(\boldsymbol{s}) = \boldsymbol{C}\boldsymbol{s} + \boldsymbol{d}, \tag{3.38}$$

with $\boldsymbol{a} \in \mathbb{R}^K, b \in \mathbb{R}, \boldsymbol{C} \in \mathbb{R}^{M\times K}$, and $\boldsymbol{d} \in \mathbb{R}^M$. Moreover, from the monotonicity of u and $\boldsymbol{h}$, it follows that

$$\boldsymbol{a} \in \mathbb{R}^K_+, \tag{3.39}$$

$$\boldsymbol{C}^{\mathrm{T}}\boldsymbol{e}_m \in \mathbb{R}^K_+. \tag{3.40}$$

The rate space problem (3.4) is equivalent to

$$\max_{s} \boldsymbol{a}^{\mathrm{T}}\boldsymbol{s} \quad \text{s.t.} \quad \boldsymbol{C}\boldsymbol{s}+\boldsymbol{d} \geq \boldsymbol{0}, \boldsymbol{s} \in \mathcal{C}. \tag{3.41}$$

The objective function in Problem (3.41) is already a weighted sum of the link rates. Accordingly, it turns out that dualizing the constraint $\boldsymbol{C}\boldsymbol{s}+\boldsymbol{d} \geq \boldsymbol{0}$ results in a dual function that is the maximum of a weighted sum rate maximization problem.

Again, a primal solution can be obtained from the solution of the dual problem if strong duality holds. Following the discussion in Sect. 3.3.1, strong duality is established by verifying that Slater's Condition is fulfilled. Accordingly, strong duality holds for Problem (3.41) if there exists $\boldsymbol{s} \in \mathcal{C}$ such that

$$\boldsymbol{C}\boldsymbol{s}+\boldsymbol{d} \gg \boldsymbol{0}. \tag{3.42}$$

Again, a feasibility problem is solved to show that Slater's condition holds. The feasibility problem is given by

$$\max_{\boldsymbol{s}\in\mathcal{C}} h(\boldsymbol{s}) \quad \text{s.t.} \quad \boldsymbol{s} \in \mathcal{C}, \tag{3.43}$$

with

$$h(\boldsymbol{s}) = \min_{m} \boldsymbol{e}_m^{\mathrm{T}}\boldsymbol{C}\boldsymbol{s} + d_m.$$

Let $\check{\boldsymbol{s}}$ be a maximizer of Problem (3.43). If $h(\check{\boldsymbol{s}}) > 0$, then Condition (3.42) is fulfilled, and strong duality holds for Problem (3.41). Problem (3.43) can be solved by considering the equivalent problem

$$\max_{(\boldsymbol{s},z)} z \quad \text{s.t.} \quad \boldsymbol{C}\boldsymbol{s}+\boldsymbol{d}-\boldsymbol{1}z \geq \boldsymbol{0}, \boldsymbol{s} \in \mathcal{C}, z \in \mathbb{R}. \tag{3.44}$$

Problem (3.44) is solved by Lagrange Duality, after showing that strong duality holds, which, in this case, is straightforward. Let $(\check{\boldsymbol{s}}, \check{z})$ be a maximizer of Problem (3.44). Then $\check{\boldsymbol{s}}$ is also a maximizer of Problem (3.43) and $h(\check{\boldsymbol{s}}) = \check{z}$. From the compactness of $\mathcal{C}$, it follows that $\check{z}$ is finite. Thus, in order to show that strong duality holds for Problem (3.44), it is sufficient to prove the existence of a feasible $(\boldsymbol{s}, z)$ such that

$$\boldsymbol{C}\boldsymbol{s}+\boldsymbol{d}-\boldsymbol{1}z \gg \boldsymbol{0}. \tag{3.45}$$

One such point is given by $(\boldsymbol{s}, z) = (\boldsymbol{0}, -1 + \min_m d_m)$:

$$\boldsymbol{C}\boldsymbol{0}+\boldsymbol{d}-\boldsymbol{1}(-1+\min_{m} d_m) = \boldsymbol{1}+\boldsymbol{d}-\boldsymbol{1}\min_{m} d_m \geq \boldsymbol{1} \gg \boldsymbol{0}.$$

Accordingly, Problem (3.44) can be solved by Lagrange duality. The dual function is given by

$$q(\boldsymbol{\lambda}) = \sup_{z\in\mathbb{R}}(1 - \boldsymbol{\lambda}^{\mathrm{T}}\mathbf{1})z + \max_{\boldsymbol{s}\in\mathcal{C}} \boldsymbol{\lambda}^{\mathrm{T}}\boldsymbol{C}\boldsymbol{s} + \boldsymbol{\lambda}^{\mathrm{T}}\boldsymbol{d}$$

$$= \begin{cases} \infty, & \boldsymbol{\lambda}^{\mathrm{T}}\mathbf{1} \neq 1, \\ \max_{\boldsymbol{s}\in\mathcal{C}} \boldsymbol{\lambda}^{\mathrm{T}}\boldsymbol{C}\boldsymbol{s} + \boldsymbol{\lambda}^{\mathrm{T}}\boldsymbol{d}, & \boldsymbol{\lambda}^{\mathrm{T}}\mathbf{1} = 1. \end{cases} \tag{3.46}$$

According to (3.46), the dual solution is obtained by solving

$$\min_{\boldsymbol{\lambda}\geq\mathbf{0}} q(\boldsymbol{\lambda}) \quad \text{s.t.} \quad \boldsymbol{\lambda}^{\mathrm{T}}\mathbf{1} = 1. \tag{3.47}$$

Let $\check{\boldsymbol{\lambda}}$ denote a minimizer of Problem (3.47). Due to the fact that strong duality holds for Problem (3.44),

$$h(\check{\boldsymbol{s}}) = \check{z} = q(\check{\boldsymbol{\lambda}}).$$

As only the maximum of Problem (3.47) is of interest to decide whether strong duality holds for Problem (3.41), there is no need to perform a primal recovery to obtain $\check{\boldsymbol{s}}$.

In the following, it is assumed that $h(\check{\boldsymbol{s}}) > 0$, i.e., strong duality holds for Problem (3.41). Under this assumption, Problem (3.41) can be solved by Lagrange duality. The Lagrangian is given by

$$L(\boldsymbol{s}, \boldsymbol{\lambda}) = \boldsymbol{a}^{\mathrm{T}}\boldsymbol{s} + \boldsymbol{\lambda}^{\mathrm{T}}(\boldsymbol{C}\boldsymbol{s} + \boldsymbol{d}), \tag{3.48}$$

and the dual function $q : \mathbb{R}_{+}^{M} \rightarrow \mathbb{R}$ follows as

$$q(\boldsymbol{\lambda}) = \max_{\boldsymbol{s}\in\mathcal{C}} L(\boldsymbol{s}, \boldsymbol{\lambda}) = \max_{\boldsymbol{s}\in\mathcal{C}}(\boldsymbol{a}^{\mathrm{T}} + \boldsymbol{\lambda}^{\mathrm{T}}\boldsymbol{C})\boldsymbol{s} + \boldsymbol{\lambda}^{\mathrm{T}}\boldsymbol{d}. \tag{3.49}$$

According to (3.49), evaluating the dual function at $\boldsymbol{\lambda}$ basically corresponds to solving a weighted sum rate maximization problem with weight vector

$$\boldsymbol{\mu} = \boldsymbol{a} + \boldsymbol{C}^{\mathrm{T}}\boldsymbol{\lambda} \in \mathbb{R}_{+}^{K}. \tag{3.50}$$

Let $\bar{\mathcal{C}}(\boldsymbol{\mu})$ denote the set of maximizers of the weighted sum rate for a given weight $\boldsymbol{\mu}$:

$$\bar{\mathcal{C}}(\boldsymbol{\mu}) = \operatorname*{argmax}_{\boldsymbol{s}\in\mathcal{C}} \boldsymbol{\mu}^{\mathrm{T}}\boldsymbol{s}.$$

The subdifferential of q at $\boldsymbol{\lambda}$ is given by

$$\partial q(\boldsymbol{\lambda}) = \left\{\boldsymbol{C}\boldsymbol{s} + \boldsymbol{d} : \boldsymbol{s} \in \bar{\mathcal{C}}(\boldsymbol{a} + \boldsymbol{C}^{\mathrm{T}}\boldsymbol{\lambda})\right\}. \tag{3.51}$$

In other words, any maximizer of the weighted sum rate maximization for weight $\boldsymbol{\mu} = \boldsymbol{a} + \boldsymbol{C}^{\mathrm{T}}\boldsymbol{\lambda}$ yields a subgradient of q at $\boldsymbol{\lambda}$. Being able to compute $q(\boldsymbol{\lambda})$ and to compute a subgradient of q at $\boldsymbol{\lambda}$, the dual problem

$$\min_{\boldsymbol{\lambda} \geq \boldsymbol{0}} q(\boldsymbol{\lambda})$$

can be solved by any of the methods for nondifferentiable convex optimization discussed in Sect. 3.3.1. Analogous to Sect. 3.3.1, a compact set $\Lambda \subset \mathbb{R}_{+}^{M}$ with $\boldsymbol{\lambda}^{*} \in \Lambda$ is given by

$$\Lambda = \left\{ \boldsymbol{\lambda} \in \mathbb{R}^{M} : 0 \leq \lambda_{m} \leq \frac{q(\boldsymbol{1}) - \boldsymbol{a}^{\mathrm{T}}\boldsymbol{s}}{h(\check{\boldsymbol{s}})} \right\}.$$

3.3.5 Lagrange Duality and Nonconvexities

If strong duality holds, Lagrange duality can be used to compute a globally optimal rate vector. Throughout this work, a basic requirement for establishing strong duality is the convexity of the rate space problem, i.e., concavity of both u and $\boldsymbol{h}$ and convexity of $\mathcal{C}$. However, convexity of the rate space problem is not necessary (nor sufficient, cf. Sect. 3.3.3) for strong duality to hold. In other words, there exist nonconvex rate space problems for which a Lagrange duality-based approach can find the optimal rate vector. Nevertheless, without the basic convexity assumptions, it is in general only possible to establish strong duality for a particular solution *after* solving the rate space problem. Moreover, the stronger the nonconvexity of the problem, the more likely it is that the rate vector obtained by Lagrange duality is not the optimal solution. In contrast, if Slater's condition is fulfilled, it is known beforehand that strong duality holds.

In this section, the implications of applying Lagrange duality to a nonconvex rate space problem are investigated. Two sources of nonconvexity are investigated: Nonconvexity in the utility model, i.e., u or $\boldsymbol{h}$ being nonconcave, and nonconvexity at the physical layer, i.e., $\mathcal{R}$ being nonconvex. The following assumption is made: Despite their nonconvexity, the dual subproblems (3.14) and (3.15) can be solved to global optimality. Based on this assumption, it is possible to make statements about the properties of the solution obtained by a dual method. However, using Lagrange duality for obtaining a solution clearly only makes sense if solving the dual subproblems to global optimality is significantly simpler than solving the rate space problem itself.

First, assume that the rate region $\mathcal{R}$ is nonconvex, while both u and $\boldsymbol{h}$ are concave. Lagrange duality is applied as described in Sect. 3.3.1, with all occurrences of $\mathcal{C}$ replaced by $\mathcal{R}$. Clearly, nonconvexity at the physical layer does not affect the utility layer subproblem (3.14). Moreover, in the physical layer subproblem (3.15), $\mathcal{R}$ can be replaced by its convex hull $\mathcal{C}$, and vice-versa, without affecting the dual solution. In other words, the dual problem is not affected by nonconvexity of the rate region. At this point, it becomes obvious that by using Lagrange duality, $\mathcal{R}$ is implicitly replaced by its convex hull $\mathcal{C}$. As a result, nonconvexity of $\mathcal{R}$ only affects the primal solution. If strong duality holds, a primal solution can be obtained by primal recovery. In case of nonconvex $\mathcal{R}$, it is not clear whether strong duality holds. Still, primal recovery can be used to obtain a rate vector $\hat{\boldsymbol{s}}$, which represents a candidate solution. Let $\boldsymbol{s}^*$ denote the (unknown) globally optimal rate vector. As using Lagrange duality implicitly replaces $\mathcal{R}$ by its convex hull $\mathcal{C}$, $\hat{\boldsymbol{s}}$ is the solution of the rate space problem

$$\max_{\boldsymbol{s} \in \mathcal{C}} u(\boldsymbol{s}) \quad \text{s.t.} \quad \boldsymbol{h}(\boldsymbol{s}) \geq \mathbf{0}.$$

Hence, $\hat{\boldsymbol{s}} \in \mathcal{C}$. Now, if $\hat{\boldsymbol{s}} \in \mathcal{R}$, then $\hat{\boldsymbol{s}}$ is the globally optimal rate vector, and strong duality holds. On the other hand, if $\hat{\boldsymbol{s}} \notin \mathcal{R}$, the dual approach returned an infeasible solution, and a means for obtaining a feasible solution $\boldsymbol{s}'$ from $\hat{\boldsymbol{s}}$ is needed. If such a solution $\boldsymbol{s}'$ can be computed, its quality can be assessed based on weak duality:

$$q(\boldsymbol{\lambda}^*) = u(\hat{\boldsymbol{s}}) \geq u(\boldsymbol{s}^*) \geq u(\boldsymbol{s}'),$$

where $\boldsymbol{\lambda}^*$ is an optimal value of the dual variable, cf. Sect. 3.3.1. If the difference $u(\hat{\boldsymbol{s}}) - u(\boldsymbol{s}')$ is small, then the optimality gap $u(\boldsymbol{s}^*) - u(\boldsymbol{s}')$ is small and $\boldsymbol{s}'$ represents a good solution. However, the difference $u(\hat{\boldsymbol{s}}) - u(\boldsymbol{s}')$ cannot be smaller than the duality gap $u(\hat{\boldsymbol{s}}) - u(\boldsymbol{s}^*)$, and the value of the duality gap is not known. Hence, if there is a significant duality gap then it is impossible to detect good solutions $\boldsymbol{s}'$.

Next, assume that the rate region is convex, while u or $\boldsymbol{h}$ are nonconcave. Again, in order to emphasize convexity of the rate region, the symbol $\mathcal{C}$ is used to denote the rate region. Recall that the utility layer subproblem is given by

$$\max_{\boldsymbol{z} \in \mathcal{Z}} u(\boldsymbol{z}) - \boldsymbol{\lambda}^{\mathrm{T}} \boldsymbol{z} \quad \text{s.t.} \quad \boldsymbol{h}(\boldsymbol{z}) \geq \mathbf{0}$$

Define two sets $\mathcal{U}$ and $\mathcal{H}'$ as follows:

$$\mathcal{U} = \{(\boldsymbol{z}, \mu) : \mu \leq u(\boldsymbol{z}), \boldsymbol{z} \in \mathcal{Z}, \mu \in \mathbb{R}\},$$
$$\mathcal{H}' = \left\{(\boldsymbol{z}, \mu) : \boldsymbol{h}(\boldsymbol{z}) \geq \mathbf{0}, \boldsymbol{z} \in \mathbb{R}_+^K, \mu \in \mathbb{R}\right\}.$$

The set $\mathcal{U}$ is the hypograph of u restricted to $\mathcal{Z}$. Using the sets $\mathcal{U}$ and $\mathcal{H}'$, the utility layer subproblem can be written as

$$\max_{(\boldsymbol{z},\mu)} \boldsymbol{w}^{\mathrm{T}}(\boldsymbol{z}, \mu) \quad \text{s.t.} \quad (\boldsymbol{z}, \mu) \in \mathcal{U} \cap \mathcal{H}', \tag{3.52}$$

with $\boldsymbol{w} = (-\boldsymbol{\lambda}, 1)$. Due to the fact that the objective in Problem (3.52) is linear, the set $\mathcal{U} \cap \mathcal{H}'$ can be replaced by its convex hull without changing the optimum, cf. Corollary A.5.9. Now let $\tilde{u}$ and $\tilde{\boldsymbol{h}}$ be two concave functions such that

$$\tilde{\mathcal{U}} \cap \tilde{\mathcal{H}}' = \operatorname{conv}\left(\mathcal{U} \cap \mathcal{H}'\right), \tag{3.53}$$

where

$$\tilde{\mathcal{U}} = \{(\boldsymbol{z}, \mu) : \mu \leq \tilde{u}(\boldsymbol{z}), \boldsymbol{z} \in \mathcal{Z}, \mu \in \mathbb{R}\},$$
$$\tilde{\mathcal{H}}' = \left\{(\boldsymbol{z}, \mu) : \tilde{\boldsymbol{h}}(\boldsymbol{z}) \geq \boldsymbol{0}, \boldsymbol{z} \in \mathbb{R}_+^K, \mu \in \mathbb{R}\right\}.$$

Moreover, assume that strong duality holds for the problem

$$\max_{(\boldsymbol{s},\boldsymbol{z})} \tilde{u}(\boldsymbol{z}) \quad \text{s.t.} \quad \tilde{\boldsymbol{h}}(\boldsymbol{z}) \geq \boldsymbol{0}, \boldsymbol{s} - \boldsymbol{z} \geq \boldsymbol{0}, (\boldsymbol{s}, \boldsymbol{z}) \in \mathcal{C} \times \mathcal{Z}. \tag{3.54}$$

By (3.53), the original rate space problem (3.9) and Problem (3.54) have the same dual function. Let $\boldsymbol{\lambda}^*$ denote a minimizer of the dual problem. Moreover, let $(\hat{\boldsymbol{s}}, \hat{\boldsymbol{z}})$ denote the solution obtained by primal recovery. Finally, let $\boldsymbol{s}^*$ denote the optimal solution of the original rate space problem (3.9). By assumption, strong duality holds for Problem (3.54), thus $(\hat{\boldsymbol{s}}, \hat{\boldsymbol{z}})$ is an optimal solution of Problem (3.54). Furthermore,

$$q(\boldsymbol{\lambda}^*) = \tilde{u}(\hat{\boldsymbol{s}}) \geq u(\boldsymbol{s}^*) \geq u(\hat{\boldsymbol{s}}).$$

Hence, if $q(\boldsymbol{\lambda}^*) = u(\hat{\boldsymbol{s}})$ and $\boldsymbol{h}(\hat{\boldsymbol{s}}) \geq \boldsymbol{0}$, then $\hat{\boldsymbol{s}}$ is an optimal solution of the rate space problem $(u, h, \mathcal{C})$. Note that the function $\tilde{u}$ need not be known, it is sufficient to have $q(\boldsymbol{\lambda}^*) = u(\hat{\boldsymbol{s}})$ and $\hat{\boldsymbol{s}}$ feasible, which implies that strong duality holds despite the nonconcavity of u or h. If $\hat{\boldsymbol{s}}$ is feasible, but $q(\boldsymbol{\lambda}^*) > u(\hat{\boldsymbol{s}})$, then the difference $q(\boldsymbol{\lambda}^*) - u(\hat{\boldsymbol{s}})$ can be used to assess the quality of the solution $\hat{\boldsymbol{s}}$, provided the (unknown) duality gap $q(\boldsymbol{\lambda}^*) - u(\boldsymbol{s}^*)$ is sufficiently small. If $\hat{\boldsymbol{s}}$ is infeasible, a method to compute a feasible solution $\boldsymbol{s}'$ is required.

In case of $\boldsymbol{h}(\boldsymbol{0}) \geq \boldsymbol{0}$ (i.e., the constraint $\boldsymbol{h}(\boldsymbol{z}) \geq \boldsymbol{0}$ can be removed), a simple interpretation of the function $\tilde{u}$ is possible: In this case,

$$\operatorname{conv}\left(\mathcal{U} \cap \mathcal{H}'\right) = \operatorname{conv}\left(\mathcal{U}\right),$$

showing that $\tilde{u}$ is the concave envelope of u (on $\mathcal{Z}$). Moreover, $\hat{\boldsymbol{s}}$ is always feasible. The duality gap is given by the difference between concave envelope and original function at $\hat{\boldsymbol{s}}$. Accordingly, strong duality is obtained if concave envelope and original function coincide at the optimal rate vector.

3.3.6 Related Work

Lagrange duality has been a popular tool for solving concave utility maximization problems since the publication of the seminal work [31], see, e.g., [14, 27] and references therein. Initially used for rate-control in wired networks with fixed link-capacities, an application of Lagrange duality in wireless settings is straightforward, provided the rate region is convex and a method for solving a weighted sum-rate maximization problem is available. As an example, the capacity region of the MIMO broadcast channel is convex, and efficient methods for solving weighted sum rate maximization problems are available, cf. Sect. 4.2. Accordingly, Lagrange duality can be used to solve a manifold of problems in the MIMO broadcast channel: In [32], a utility maximization problem with MIMO broadcast channels at the physical layer is solved, for general concave utility functions, while in [33], the utility function is a weighted sum of logarithms, yielding the so-called weighted proportionally fair rate vector. Furthermore, [34] computes the max–min rate vector, while [35] maximizes sum-rate under minimum rate constraints.

A few works have considered the application of Lagrange duality in a network with fixed link capacities and *nonconcave* utilities. In [36], a standard subgradient-based dual method is used to compute a rate allocation in a network with sources that use sigmoidal utility functions.[2] The authors in [36] observe an oscillation in the source rates, and propose a "self-regulating" heuristic to eliminate the oscillating behavior. Importantly, the source rates (i.e., the primal solution) are not obtained by primal recovery. Instead, at each iteration, a maximizer of the utility layer problem (3.14) is taken as the current vector of source rates. For a given value of the dual variable, the set of maximizers may not be unique. As a result, if the optimal value of the dual variable corresponds to a value for which the set of maximizers is not unique, choosing the value of the primal variable as in [36] leads to oscillating behavior of the primal variable as the dual iterates get close to the optimal value. A sufficient condition for the set of maximizers to be a singleton is the strict concavity of u. As discussed in Sect. 3.3.5, solving a problem with nonconcave utility function via Lagrange duality is equivalent to replacing the utility function by its concave envelope. The concave envelope of a sigmoidal function is not strictly concave, hence the oscillating behavior in [36]. It is interesting to note that any concave, but not strictly concave utility function would lead to exactly the same oscillations. The "self-regulating" heuristic proposed in [36] is unnecessary—instead, it is sufficient to perform a primal recovery by simply averaging over the maximizers of the application layer problem, see, e.g., [37].

Closely related to [36] is [38], which considers the same setup as [36] and provides conditions under which a standard Lagrange duality based approach can be used to compute the globally optimal rate allocation, despite the nonconcavity of the utility functions. In Sect. 3.3.5, such a condition is obtained by considering the concave envelope of the utility function: obviously, the rate allocation is globally optimal if nonconcave utility function and its concave envelope coincide at the optimal solution. It turns out that the conditions presented in [38] are nothing else than conditions for this coincidence.

---

[2]For an example of a sigmoidal utility function, see Sect. 5.2.

A fundamental problem when applying Lagrange duality to nonconvex problems is the fact that the value of the duality gap is generally unknown, cf. Sect. 3.3.5. As a result, it is often not possible to assess the quality of heuristic solutions based on the dual upper bound. In [39], Lagrange duality is applied to the parameter space formulation of a utility-based power control problem. By exploiting the specific structure of the problem, the authors of [39] are able to obtain a nontrivial upper bound on the loss in terms of overall utility by using their duality-based heuristic instead of the globally optimal solution.

3.4 Monotonic Optimization

Section 3.3 discusses solution approaches for a utility maximization problem $(u, h, \boldsymbol{r}, \mathcal{X}_1)$ under the assumption that both u and h are concave and that the rate region $\mathcal{R} = \boldsymbol{r}(\mathcal{X}_1)$ is convex. While convexity of the rate region can be obtained by providing a time sharing mode at the physical layer, cf. Sect. 2.3.5, concavity of u and h represents a nontrivial requirement. Moreover, the Lagrange duality-based approach relies on a method to maximize a weighted sum of rates at the physical layer. If an efficient algorithm for maximizing a weighted sum of rates is not available for the physical layer model under consideration, Lagrange duality-based approaches are of limited value. As an example, for the MISO BC with linear precoding, there is no efficient solution for weighted sum rate maximization. On the other hand, a problem that can be efficiently solved is computing the intersection of a line and the Pareto boundary of the rate region *without time sharing*, cf. Sect. 4.1.3. The last observation provides a clear motivation for also considering nonconvex rate regions $\mathcal{R}$, and this general approach is justified in Sect. 6.1, where it is shown that time sharing can actually lead to significantly increased complexity of the solution method.

This section presents a solution method that does not rely on any of the additional assumptions made by the Lagrange duality-based approach. In contrast, the monotonicity structure implied by the definition of a utility maximization problem is sufficient for the solution approach presented in this section.

In order to apply Lagrange duality, convexity at the rate level is required, i.e., u and h are concave functions of the rate, and the rate region is convex. The key to exploiting this convexity structure is to address the rate space problem (3.4) instead of the original parameter space problem (2.30), as discussed in Sects. 3.2 and 3.3. By definition, any utility maximization problem offers a monotonicity structure at the rate level: the functions u and h are increasing. Moreover, the rate region $\mathcal{R}$ associated with the physical layer $(\boldsymbol{r}, \mathcal{X}_1)$ is compact. Accordingly, the rate space problem (3.4) corresponds to the maximization of an increasing function over a compact set under an increasing constraint. Optimization problems of this type fall within the framework of *monotonic optimization* [24]. Thus, analogous to the Lagrange duality-based approach, the key to exploiting the monotonicity structure is again to address the rate space problem (3.4). The rate level monotonicity structure alone, however, is not sufficient to solve the rate space problem. Like in the case of Lagrange duality, where the availability of a method to maximize a weighted sum of rates represents an elementary requirement, again a method is required to compute specific points in $\mathcal{R}$. In case of monotonic optimization, this elementary method is not weighted sum rate maximization, but computing the intersection of a line and the weak Pareto boundary of the rate region $\mathcal{R}$.

3.4.1 Polyblock Algorithm

The so-called polyblock algorithm [24, 40] is a generic algorithm for solving monotonic optimization problems to global optimality. Accordingly, it is used to solve the rate space problem (3.4) in the general case that the convexity requirements for a Lagrange duality-based approach are not satisfied.

Let $(u, h, \boldsymbol{r}, \mathcal{X}_1)$ denote a utility maximization problem according to Definition 2.2.1. In addition, it is assumed that the rate region $\mathcal{R} = \boldsymbol{r}(\mathcal{X}_1)$ is proper and that the domain of h is $\mathbb{R}_+^K$. Let

$$\mathcal{H} = \left\{ \boldsymbol{s} \in \mathbb{R}_+^K : h(\boldsymbol{s}) \geq 0 \right\}.$$

According to Proposition 2.3.1, it is sufficient to maximize u over the feasible points on the Pareto boundary. A similar result is given in [24], where it is shown that the maximum is attained on the weak Pareto boundary. With Proposition 2.3.1, the latter result follows immediately from the inclusion

$$\operatorname{par}(\mathcal{R}) \subseteq \operatorname{wpar}(\mathcal{R}).$$

As a consequence of Proposition 2.3.1, in the rate space problem (3.4), $\mathcal{R}$ can be replaced by $\operatorname{par}(\mathcal{R})$. In general, replacing $\mathcal{R}$ by its Pareto boundary does not simplify the optimization problem. There exists a family of compact sets, however, for which the Pareto set has a very simple structure. These sets are the so-called polyblocks, which are given by the union of a finite number of hyperrectangles (or boxes). Given a finite set $\mathcal{V} \subset \mathbb{R}_+^K$, a polyblock $\mathcal{P}(\mathcal{V})$ is defined as

$$\mathcal{P}(\mathcal{V}) = \bigcup_{\boldsymbol{v} \in \mathcal{V}} \{\boldsymbol{x} \in \mathbb{R}_+^K, \boldsymbol{x} \leq \boldsymbol{v}\} = \operatorname{comp}(\mathcal{V}).$$

The set $\mathcal{V}$ contains the vertices of the polyblock $\mathcal{P}(\mathcal{V})$. Clearly,

$$\operatorname{par}(\mathcal{P}(\mathcal{V})) \subseteq \mathcal{V}. \tag{3.55}$$

Vertex sets for which equality holds in (3.55) are denoted as proper [24]. According to Inequality (3.55) and Proposition 2.3.1, if $\mathcal{R}$ is a polyblock, the rate space

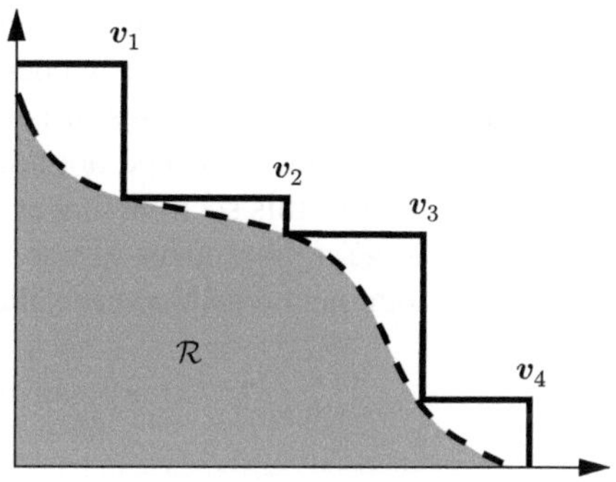

Fig. 3.1 Outer approximation of a proper set by a polyblock

problem (3.4) can be solved to global optimality by performing an exhaustive search over the finite vertex set $\mathcal{V}$.

In general, though, the rate region $\mathcal{R}$ is not a polyblock. The idea underlying the polyblock algorithm is to iteratively generate a sequence of vertex sets $\mathcal{V}^{(0)}, \ldots, \mathcal{V}^{(N)}$ such that each polyblock $\mathcal{P}(\mathcal{V}^{(n)})$ is an outer approximation of $\mathcal{R}$,

$$\mathcal{R} \subseteq \mathcal{P}(\mathcal{V}^{(n)}), \tag{3.56}$$

and the quality of the approximation is improved in each iteration,

$$\mathcal{P}(\mathcal{V}^{(n+1)}) \subset \mathcal{P}(\mathcal{V}^{(n)}), \tag{3.57}$$

until the upper bound on the maximal utility value obtained by maximizing over the polyblock $\mathcal{P}(\mathcal{V}^{(n)})$ is sufficiently tight. Clearly, (3.57) can only be satisfied as long as $\mathcal{R} \subset \mathcal{P}(\mathcal{V}^{(n)})$. The approximation of a proper set by a polyblock is illustrated in Fig. 3.1.

In [24], the polyblock algorithm is derived under the assumption[3] that

$$\mathcal{H} \subset \mathbb{R}^K_{++}. \tag{3.59}$$

In the context of this work, however, Condition (3.59) is overly demanding. For example, it excludes the important case that the constraint $h(\boldsymbol{s}) \geq 0$ is not present,

---

[3]In fact, [24] starts with the assumption that

$$\mathcal{R} \cap \mathcal{H} \subset \mathbb{R}^K_{++}. \tag{3.58}$$

It is shown in [24] that if Condition (3.58) holds, there exists a compact set $\tilde{\mathcal{H}} \subset \mathbb{R}^K_{++}$ such that

$$\mathcal{R} \cap \mathcal{H} = \mathcal{R} \cap \tilde{\mathcal{H}}.$$

In other words, if Condition (3.58) holds, $\mathcal{H}$ can be replaced by a set that satisfies Condition (3.59). Based on this result, in [24], the polyblock algorithm is derived under the assumption that $\mathcal{H}$ satisfies Condition (3.59).

which is equivalent to $\mathcal{H} = \mathbb{R}_+^K$. Moreover, for an arbitrary utility maximization problem $(u, h, \mathcal{R})$, it may well happen that at least one link has zero rate in the optimum, i.e., $\boldsymbol{s}^* \notin \mathbb{R}_{++}^K$. Thus, it is desirable to extend the polyblock algorithm such that Condition (3.59) need not hold. This extension is provided in Sect. 3.4.3. In the remainder of this section, however, it is assumed that Condition (3.59) holds.

Let $\boldsymbol{s}^*$ denote a maximizer of the rate space problem (3.4). Given a vertex set $\mathcal{V}^{(n)}$, let $\overline{\boldsymbol{s}}^{(n)}$ denote a feasible vertex that maximizes utility, i.e.,

$$\overline{\boldsymbol{s}}^{(n)} \in \underset{\boldsymbol{s}\in\mathcal{V}^{(n)}}{\operatorname{argmax}}\, u(\boldsymbol{s}) \quad \text{s.t.} \quad h(\boldsymbol{s}) \geq \mathbf{0}.$$

By (3.56), $\overline{\boldsymbol{s}}^{(n)}$ provides an upper bound $\overline{u}^{(n)} = u(\overline{\boldsymbol{s}}^{(n)})$ on the maximal utility value, i.e.,

$$u(\boldsymbol{s}^*) \leq \overline{u}^{(n)}. \tag{3.60}$$

Moreover, by (3.57),

$$\overline{u}^{(n+1)} \leq \overline{u}^{(n)}.$$

If $\overline{\boldsymbol{s}}^{(n)} \in \mathcal{R}$, an optimal rate vector is found. In general, though, $\overline{\boldsymbol{s}}^{(n)} \notin \mathcal{R}$, and a new vertex set $\mathcal{V}^{(n+1)}$ satisfying (3.57) is generated. To this end, the point $\overline{\boldsymbol{s}}^{(n)}$ is first projected on the weak Pareto boundary of $\mathcal{R}$. By Condition (3.59), $\overline{\boldsymbol{s}}^{(n)} \in \mathbb{R}_{++}^K$. Let $\pi : \mathbb{R}_{++}^K \to \mathbb{R}_+^K$ denote a projection rule satisfying

$$\pi(\boldsymbol{s}) \in \operatorname{wpar}(\mathcal{R}), \forall \boldsymbol{s} \in \mathbb{R}_{++}^K, \tag{3.61}$$

$$\pi(\boldsymbol{s}) = \boldsymbol{s}, \forall \boldsymbol{s} \in \operatorname{wpar}(\mathcal{R}), \tag{3.62}$$

$$\pi(\boldsymbol{s}) \ll \boldsymbol{s}, \forall \boldsymbol{s} \in \mathbb{R}_{++}^K \setminus \mathcal{R}. \tag{3.63}$$

Define

$$\boldsymbol{s}^{(n)} = \pi(\overline{\boldsymbol{s}}^{(n)}).$$

By (3.61), if $\boldsymbol{s}^{(n)} \in \mathcal{H}$, then $\boldsymbol{s}^{(n)}$ provides a lower bound on the maximal utility value,

$$\boldsymbol{s}^{(n)} \in \operatorname{wpar}(\mathcal{R}) \cap \mathcal{H} \implies \boldsymbol{s}^{(n)} \in \mathcal{R} \cap \mathcal{H} \implies u(\boldsymbol{s}^{(n)}) \leq u(\boldsymbol{s}^*).$$

Let $\underline{u}^{(n)}$ denote the best lower bound obtained so far, i.e.,

$$\underline{u}^{(n)} = \sup\left\{u(\boldsymbol{s}) : \boldsymbol{s} \in \left\{\boldsymbol{s}^{(1)}, \ldots, \boldsymbol{s}^{(n)}\right\} \cap \mathcal{H}\right\}. \tag{3.64}$$

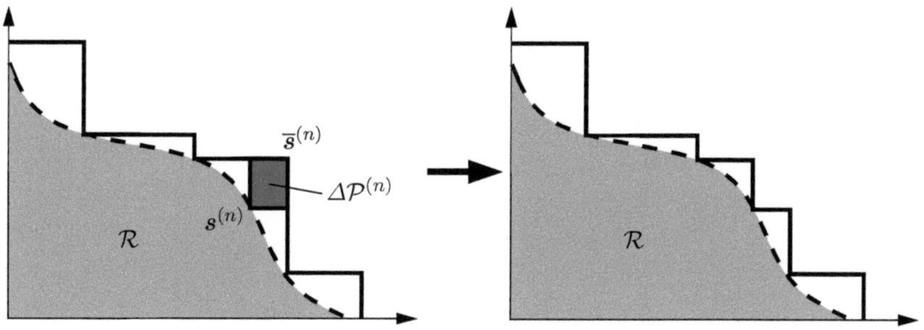

Fig. 3.2 The polyblock approximation is refined by cutting away the shifted cone $\{s^{(n)}\} + \mathbb{R}^K_{++}$

Given a point $s^{(n)}$ on the weak Pareto boundary of $\mathcal{R}$, the polyblock approximation can be refined. This process is illustrated in Fig. 3.2. From $s^{(n)} \in \mathrm{wpar}(\mathcal{R})$, it follows that

$$\mathcal{R} \cap (\{s^{(n)}\} + \mathbb{R}^K_{++}) = \emptyset. \tag{3.65}$$

As a result, the set

$$\Delta\mathcal{P}^{(n)} = (\{s^{(n)}\} + \mathbb{R}^K_{++}) \cap \mathcal{P}(\mathcal{V}^{(n)}) = \bigcup_{v \in \mathcal{V}^{(n)}} \{s \in \mathbb{R}^K_+ : s^{(n)} \ll s \leq v\}$$

can be removed from $\mathcal{P}(\mathcal{V}^{(n)})$ without removing any feasible rate vectors:

$$\mathcal{R} \cap \Delta\mathcal{P}^{(n)} = \emptyset. \tag{3.66}$$

Consequently, the new vertex set $\mathcal{V}^{(n+1)}$ is chosen such that

$$\mathcal{P}(\mathcal{V}^{(n+1)}) = \mathcal{P}(\mathcal{V}^{(n)}) \setminus \Delta\mathcal{P}^{(n)}.$$

As shown in [24], if the projection rule fulfills (3.63), the set $\mathcal{P}(\mathcal{V}^{(n)}) \setminus \Delta\mathcal{P}^{(n)}$ is a polyblock, thus such a vertex set exists. Moreover, by (3.66), the polyblock $\mathcal{P}(\mathcal{V}^{(n+1)})$ is again an outer approximation, i.e., it satisfies (3.56). Finally, note that

$$\overline{s}^{(n)} \in \mathcal{V}^{(n)},$$

thus

$$\Delta\mathcal{P}^{(n)} \supseteq \{s \in \mathbb{R}^K_+ : s^{(n)} \ll s \leq \overline{s}^{(n)}\},$$

and

$$\text{vol}\left(\Delta\mathcal{P}^{(n)}\right) \geq \prod_{k=1}^{K}\left(\overline{s}_k^{(n)} - s_k^{(n)}\right).$$

From (3.63) it follows that

$$\overline{s}_k^{(n)} - s_k^{(n)} > 0, \forall k,$$

thus

$$\text{vol}\left(\Delta\mathcal{P}^{(n)}\right) > 0. \tag{3.67}$$

Equation (3.67) shows that $\Delta\mathcal{P}^{(n)}$ is nonempty, thus (3.57) holds.

The algorithm terminates when

$$\overline{u}^{(n)} - \underline{u}^{(n)} \leq \epsilon,$$

where ϵ is a given tolerance. By 3.64 there exists $\ell \in \{1, \ldots, n\}$ such that

$$u(\boldsymbol{s}^{(\ell)}) = \underline{u}^{(n)}.$$

Hence, when the algorithm terminates, by (3.60) the point $\boldsymbol{s}^{(\ell)}$ is an ϵ-optimal solution:

$$u(\boldsymbol{s}^*) - u(\boldsymbol{s}^{(\ell)}) \leq \overline{u}^{(n)} - \underline{u}^{(n)} \leq \epsilon.$$

3.4.2 *Projection Rules*

The final step in the polyblock algorithm is to select a projection rule π such that the sequences $\left\{\underline{u}^{(n)}\right\}_{n=1}^{\infty}$ and $\left\{\overline{u}^{(n)}\right\}_{n=1}^{\infty}$ both converge to u^*. Two such projection rules are provided in [24].

3.4.2.1 Projection Along Ray Through Origin

The first projection rule proposed in [24] is to choose as projection point $\pi(\boldsymbol{s})$ the intersection point of the line through $\mathbf{0}$ and $\boldsymbol{s}$ and the weak Pareto boundary of $\mathcal{R}$:

$$\pi(\boldsymbol{s}) = \gamma^* \boldsymbol{s}, \tag{3.68}$$

with

$$\gamma^* = \max\left\{\gamma \in \mathbb{R} : \gamma \boldsymbol{s} \in \mathcal{R}\right\}.$$

By assumption, the rate region $\mathcal{R}$ is proper, thus, by Proposition A.4.4, the projection rule is well-defined for $\boldsymbol{s} \in \mathbb{R}^K_{++}$, and $\gamma^* \boldsymbol{s} \in \mathrm{wpar}(\mathcal{R})$. Moreover, if $\boldsymbol{s} \in \mathbb{R}^K_{++} \setminus \mathcal{R}$, then $\gamma^* < 1$, thus $\gamma^* \boldsymbol{s} \ll \boldsymbol{s}$.

3.4.2.2 Projection Along Arbitrary Direction

Let $\boldsymbol{d}$ be an arbitrary vector in $\mathbb{R}^K_{++}$. The second projection rule proposed in [24] is to choose as projection point $\pi(\boldsymbol{s})$ the intersection point between the line

$$\{\boldsymbol{s} + \gamma \boldsymbol{d}, \gamma \in \mathbb{R}\}$$

and the weak Pareto boundary of $\mathcal{R}$:

$$\pi(\boldsymbol{s}) = \boldsymbol{s} + \gamma^* \boldsymbol{d}, \tag{3.69}$$

with

$$\gamma^* = \sup \{\gamma \in \mathbb{R} : \boldsymbol{s} + \gamma \boldsymbol{d} \in \mathcal{R}\}.$$

However, for arbitrary $\boldsymbol{d} \in \mathbb{R}^K_{++}$, the set

$$\{\gamma \in \mathbb{R} : \boldsymbol{s} + \gamma \boldsymbol{d} \in \mathcal{R}\}$$

may be empty. In this case, the second projection rule is not well-defined. This possibility is not considered in [24]. In order to obtain a well-defined projection rule, define a function $\phi : \mathbb{R} \to \mathbb{R}^K_+$ as follows:

$$\phi(\gamma) = (\boldsymbol{s} + \gamma \boldsymbol{d})^+.$$

A well-defined projection rule π is then given by

$$\pi(\boldsymbol{s}) = \phi(\gamma^*), \tag{3.70}$$

where

$$\gamma^* = \max \{\gamma \in \mathbb{R} : \phi(\gamma) \in \mathcal{R}\}. \tag{3.71}$$

The function ϕ is continuous and increasing. By Proposition A.4.4, $\phi(\gamma^*) \in \mathrm{wpar}(\mathcal{R})$. In addition, if $\boldsymbol{s} \in \mathbb{R}^K_{++} \setminus \mathcal{R}$, then $\boldsymbol{d} \in \mathbb{R}^K_{++}$ implies $\phi(\gamma^*) \ll \boldsymbol{s}$.

Note that problem (3.71) is an intersection problem as defined in Sect. 2.3.2. Moreover, the projection rule defined by (3.68) is included as a special case, by letting $\boldsymbol{d} = \boldsymbol{s}$.

In [24], it is shown that using the projection rule defined in (3.69), the sequences $\{\underline{u}^{(n)}\}_1^\infty$ and $\{\overline{u}^{(n)}\}_1^\infty$ converge to the optimum, under the assumption that the projection rule is well-defined for all $\overline{\boldsymbol{s}}^{(n)}$. The convergence proof relies on the property that

$$\min_k d_k = \delta > 0.$$

Accordingly, it is possible to use a different direction $\boldsymbol{d}^{(n)}$ at each iteration, as long as

$$\min_k d_k^{(n)} \geq \delta > 0, \forall n,$$

where δ is a constant that does not depend on n. Based on this observation, it is straightforward to prove convergence for the projection rule defined in (3.70). According to (3.70), $\boldsymbol{s}^{(n)}$ can be written as

$$\boldsymbol{s}^{(n)} = \overline{\boldsymbol{s}}^{(n)} + \gamma^* \boldsymbol{d}^{(n)},$$

with an effective direction $\boldsymbol{d}^{(n)}$. Define

$$\mathcal{I}(\boldsymbol{s}) = \{k : s_k = 0\}.$$

The k-th coordinate of $\boldsymbol{d}^{(n)}$ is given by

$$d_k^{(n)} = \begin{cases} d_k, & k \notin \mathcal{I}(\boldsymbol{s}^{(n)}), \\ -\frac{\overline{s}_k^{(n)}}{\gamma^*}, & k \in \mathcal{I}(\boldsymbol{s}^{(n)}). \end{cases} \tag{3.72}$$

Note that $\overline{\boldsymbol{s}}^{(n)} \gg \boldsymbol{s}^{(n)}$, thus $\gamma^* < 0$, and $\boldsymbol{d}^{(n)} \in \mathbb{R}_{++}^K$.

Proposition 3.4.1. *There exists $\delta > 0$ such that*

$$\min_k d_k^{(n)} \geq \delta, \forall n.$$

Proof. Note that $\boldsymbol{d} \in \mathbb{R}_{++}^K$, thus it suffices to show that there exists $\delta > 0$ such that

$$-\frac{\overline{s}_k^{(n)}}{\gamma^*} \geq \delta, \forall k, n.$$

By assumption, $\mathcal{H} \subset \mathbb{R}_{++}^K$. In addition, $\mathcal{H}$ is closed, thus there exists $\boldsymbol{a} \gg \boldsymbol{0}$ such that

$$\boldsymbol{s} \geq \boldsymbol{a}, \forall \boldsymbol{s} \in \mathcal{H}. \tag{3.73}$$

Moreover, $\mathcal{P}(\mathcal{V}^{(0)})$ is bounded, and $\mathcal{P}(\mathcal{V}^{(n+1)}) \subset \mathcal{P}(\mathcal{V}^{(n)})$, thus there exists $\boldsymbol{b} \in \mathbb{R}_+^K$ such that

$$\boldsymbol{b} \geq \boldsymbol{s}, \forall \boldsymbol{s} \in \mathcal{P}(\mathcal{V}^{(n)}), \forall n. \tag{3.74}$$

Let

$$\gamma' = -\max_q \frac{b_q}{d_q}.$$

With (3.74),

$$\max_q \frac{b_q}{d_q} \geq \max_q \frac{\bar{s}_q^{(n)}}{d_q} \geq \frac{\bar{s}_k^{(n)}}{d_k}.$$

As a result,

$$\bar{s}_k^{(n)} + \gamma' d_k \leq \bar{s}_k^{(n)} - \frac{\bar{s}_k^{(n)}}{d_k} d_k = 0,$$

thus $\phi(\gamma') = 0 \in \mathcal{R}$ and, by definition of γ^*,

$$\gamma' \leq \gamma^* < 0. \tag{3.75}$$

Note that γ' does not depend on n, while γ^* is computed at each iteration. With (3.75) and (3.73),

$$-\frac{\bar{s}_k^{(n)}}{\gamma^*} = \frac{\bar{s}_k^{(n)}}{|\gamma^*|} \geq \frac{\bar{s}_k^{(n)}}{|\gamma'|} \geq \frac{a_k}{|\gamma'|}.$$

Letting $\delta = \min_k \frac{a_k}{|\gamma'|}$ and observing that $\boldsymbol{a} \gg \boldsymbol{0}$ yields the result. □

Remark 3.4.1. Note that $\boldsymbol{d}^{(n)} \in \mathbb{R}_{++}^K$ is not sufficient to establish the result. Without further restrictions on $\boldsymbol{d}^{(n)}$, there exists a sequence $\{\boldsymbol{d}^{(n)}\}$, with $\boldsymbol{d}^{(n)} \in \mathbb{R}_{++}^K$, such that the sequence $\{\min_k d_k^{(n)}\}_{n=1}^{\infty}$ has a limit point 0, which can prevent the polyblock algorithm from converging to the global optimum. Proposition 3.4.1 shows that if the $\boldsymbol{d}^{(n)}$ are chosen according to (3.72), then 0 is not a limit point.

3.4.3 Extended Polyblock Algorithm

As discussed in Sect. 3.4.1, the polyblock algorithm is derived under the assumption that

$$\mathcal{H} \subset \mathbb{R}^K_{++}.$$

In the context of utility maximization, this assumption is overly restrictive. Fortunately, given a problem with $\mathcal{H} \not\subset \mathbb{R}^K_{++}$, an equivalent monotonic problem that satisfies assumption (3.59) can be obtained by simply shifting both $\mathcal{R}$ and $\mathcal{H}$.

Let $\boldsymbol{a} \in \mathbb{R}^K_{++}$. Given the original problem $(u, h, \mathcal{R})$, define the shifted problem as follows:

$$\tilde{u}(\boldsymbol{s}) = u((\boldsymbol{s} - \boldsymbol{a})^+), \tag{3.76}$$

$$\tilde{h}(\boldsymbol{s}) = \min\left\{h((\boldsymbol{s} - \boldsymbol{a})^+), s_1 - a_1, \ldots, s_K - a_K\right\}, \tag{3.77}$$

$$\tilde{\mathcal{R}} = \operatorname{comp}(\{\boldsymbol{a}\} + \mathcal{R}). \tag{3.78}$$

Proposition 3.4.2. *The shifted problem*

$$\max_{\boldsymbol{s} \in \tilde{\mathcal{R}}} \tilde{u}(\boldsymbol{s}) \quad \text{s.t.} \quad \tilde{h}(\boldsymbol{s}) \geq 0$$

is a monotonic optimization problem. Moreover,

$$\tilde{\mathcal{H}} = \left\{\boldsymbol{s} : \tilde{h}(\boldsymbol{s}) \geq 0\right\} \subset \mathbb{R}^K_{++},$$

i.e., the shifted problem can be solved using the polyblock algorithm.

Proof. The set $\tilde{\mathcal{R}}$ is compact and comprehensive. The functions $\tilde{u}$ and $\tilde{h}$ are increasing. Thus, the shifted problem is a monotonic optimization problem.

Let $\boldsymbol{s} \in \tilde{\mathcal{H}}$. By definition of $\tilde{h}$, this implies $\boldsymbol{s} \geq \boldsymbol{a}$, thus $\boldsymbol{s} \in \mathbb{R}^K_{++}$. □

Proposition 3.4.3. *Let $\tilde{\boldsymbol{s}}$ denote a maximizer of the shifted problem. Then $(\tilde{\boldsymbol{s}} - \boldsymbol{a})^+$ is a maximizer of the original problem.*

Proof. Observe that

$$\tilde{\mathcal{R}} = \operatorname{comp}(\{\boldsymbol{a}\} + \mathcal{R}) = \bigcup_{\boldsymbol{r} \in \mathcal{R}} \left\{\boldsymbol{s} \in \mathbb{R}^K_+ : \boldsymbol{s} \leq \boldsymbol{r} + \boldsymbol{a}\right\}.$$

Thus,

$$\operatorname{par}\left(\tilde{\mathcal{R}}\right) = \bigcup_{\boldsymbol{r} \in \operatorname{par}(\mathcal{R})} \{\boldsymbol{a} + \boldsymbol{r}\} = \{\boldsymbol{a}\} + \operatorname{par}(\mathcal{R}). \tag{3.79}$$

As a result,

$$\boldsymbol{s} - \boldsymbol{a} \geq \boldsymbol{0}, \forall \boldsymbol{s} \in \operatorname{par}\left(\tilde{\mathcal{R}}\right). \tag{3.80}$$

Using Proposition 2.3.1, (3.79), and (3.80),

$$
\begin{aligned}
u((\tilde{s}-a)^{+}) &= \tilde{u}(\tilde{s}) \\
&= \max\left\{\tilde{u}(s) : s \in \tilde{\mathcal{R}}, \tilde{h}(s) \geq 0\right\} \\
&= \max\left\{\tilde{u}(s) : s \in \operatorname{par}\left(\tilde{\mathcal{R}}\right), \tilde{h}(s) \geq 0\right\} \\
&= \max\left\{u(s-a) : s \in \operatorname{par}\left(\tilde{\mathcal{R}}\right), h(s-a) \geq 0\right\} \\
&= \max\left\{u(s) : s \in \operatorname{par}\left(\mathcal{R}\right), h(s) \geq 0\right\} \\
&= \max\left\{u(s) : s \in \mathcal{R}, h(s) \geq 0\right\}.
\end{aligned}
$$

Thus, $(\tilde{s}-a)^{+}$ is a maximizer of the original problem. □

According to Proposition 3.4.3, a utility maximization problem with $\mathcal{H} \not\subset \mathbb{R}^{K}_{++}$ can be solved by solving the shifted problem. According to Proposition 3.4.2, the shifted problem can be solved with the standard polyblock algorithm described in Sect. 3.4.1. At each iteration of the polyblock algorithm, a vertex $\overline{s}^{(n)}$ is projected on the weak Pareto boundary of $\tilde{\mathcal{R}}$, cf. Sect. 3.4.1. Using the general projection rule defined in (3.70), each projection step involves solving a problem of the following type:

$$
\max_{\gamma \in \mathbb{R}} \gamma \quad \text{s.t.} \quad \phi(\gamma) \in \tilde{\mathcal{R}}, \tag{3.81}
$$

where $\phi(\gamma) \in \mathbb{R}^{K}_{+}$. Proposition 3.4.4 shows that Problem (3.81) can be solved by solving an equivalent problem on the original rate region $\mathcal{R}$.

Proposition 3.4.4. *A vector $s \in \mathbb{R}^{K}_{+}$ is element of $\tilde{\mathcal{R}}$ if and only if the vector $(s-a)^{+}$ is element of $\mathcal{R}$.*

Proof. Let $s \in \tilde{\mathcal{R}}$. By definition of $\tilde{\mathcal{R}}$, there exists $r \in \mathcal{R}$ such that

$$
s \leq r + a.
$$

Re-arranging, applying the $(\cdot)^{+}$ operator on both sides, and noting that $(r)^{+} = r$ yields

$$
(s-a)^{+} \leq r.
$$

Thus, by comprehensiveness of $\mathcal{R}$, $(s-a)^{+} \in \mathcal{R}$.

Now, assume that $(s-a)^{+} \in \mathcal{R}$. By definition of $\tilde{\mathcal{R}}$,

$$
(s-a)^{+} \in \mathcal{R} \implies (s-a)^{+} + a \in \tilde{\mathcal{R}}. \tag{3.82}
$$

Moreover,

$$\begin{aligned} & s - a \le (s - a)^+ \\ \implies \quad & s \le (s - a)^+ + a \\ \implies \quad & s \in \tilde{\mathcal{R}}, \end{aligned}$$

where the last implication follows from (3.82), the comprehensiveness of $\tilde{\mathcal{R}}$, and $s \in \mathbb{R}_+^K$. □

Using Proposition 3.4.4, the solution of Problem (3.81) can be obtained by solving an intersection problem on the original rate region $\mathcal{R}$:

$$\begin{aligned} \gamma^* &= \max\left\{\gamma \in \mathbb{R} : \phi(\gamma) \in \tilde{\mathcal{R}}\right\} && (3.83) \\ &= \max\left\{\gamma \in \mathbb{R} : (\phi(\gamma) - a)^+ \in \mathcal{R}\right\}. && (3.84) \end{aligned}$$

Moreover, note that

$$(\phi(\gamma) - a)^+ = \left((s + \gamma d)^+ - a\right)^+ = (s - a + \gamma d)^+.$$

Thus, Problem (3.84) is just a standard intersection problem on $\mathcal{R}$ as defined in Sect. 2.3.2. To summarize, the optimal value γ^* can be obtained by solving an intersection problem on $\mathcal{R}$. The point $s^{(n)}$ on the weak Pareto boundary of $\tilde{\mathcal{R}}$ is given by

$$s^{(n)} = \phi(\gamma^*) = (\overline{s}^{(n)} + \gamma^* d)^+,$$

where $\overline{s}^{(n)}$ is a feasible vertex of the polyblock $\mathcal{P}(\mathcal{V}^{(n)})$ that maximizes utility, cf. Sect. 3.4.1. In the standard polyblock algorithm, the point $s^{(n)}$ is used to obtain a lower bound on the maximum utility value, cf. (3.64). By Proposition 3.4.4, if $s^{(n)} \in \tilde{\mathcal{R}}$, then $(s^{(n)} - a)^+ \in \mathcal{R}$. According to the definition of $\tilde{\mathcal{R}}$, this implies $(s^{(n)} - a)^+ + a \in \tilde{\mathcal{R}}$. Clearly

$$(s^{(n)} - a)^+ + a \ge s^{(n)},$$

thus an improved bound is given by

$$\underline{u}^{(n)} = \sup\left\{\tilde{u}(s) : s \in \left\{(s^{(1)} - a)^+ + a, \ldots, (s^{(n)} - a)^+ + a\right\} \cap \tilde{\mathcal{H}}\right\}. \quad (3.85)$$

3.4.4 Solving the Intersection Problem

At each iteration of the polyblock algorithm, an intersection problem has to be solved. Ideally, the physical layer provides an efficient algorithm for solving an intersection problem directly. As an example, the intersection problem can be efficiently solved in the MISO BC with linear precoding, as discussed in Sect. 4.1.3. This approach is followed in Sect. 6.1.

Proposition 5.1.1 shows that for comprehensive rate regions, the intersection problem is equivalent to a generalized max–min problem. Moreover, the generalized max–min problem is a utility maximization problem with a concave utility function, cf. Sect. 5.1.3. Accordingly, if the rate region is convex and the physical layer provides an algorithm for maximizing a weighted sum of rates, the generalized max–min problem (and thereby the intersection problem) can be solved using the methods from Sect. 3.3. This approach is followed in Sects. 6.1 and 6.2

If neither an algorithm for solving an intersection problem directly nor an algorithm for maximizing a weighted sum of rates is available, the intersection problem cannot be solved, and, as a consequence, the framework developed in this section is not applicable to the corresponding utility maximization problem.

3.4.5 Complexity of the Polyblock Algorithm

The polyblock algorithm is used to compute an ϵ-optimal solution of a nonconvex optimization problem. In order to find the global ϵ-optimum, the utility function u is repeatedly evaluated at the feasible vertices of the approximating polyblock. The only structure of the utility function that is exploited by the algorithm is the fact that u is increasing. Apart from that, the polyblock algorithm uses a black box model of the utility function: given a value $\boldsymbol{s} \in \mathbb{R}_+^K$, the black box returns a value $u(\boldsymbol{s})$. A basic measure for the complexity of the polyblock algorithm is the maximum number of iterations needed until an ϵ-optimal solution is found, as a function of the number of links K. The maximum is taken with respect to the triple $(u, h, \mathcal{R})$, i.e., the worst case complexity over the class of functions u and h and the family of proper sets $\mathcal{R}$ is considered. Note that this basic measure does not consider the computations which have to be performed in each iteration, such as solving the intersection problem and generating the new vertex set. Without further restrictions on the class of functions u, this measure is not meaningful: According to Definition 2.2.1, u belongs to the class of continuous increasing functions. Choose h and $\mathcal{R}$ such that $\mathcal{R} \cap \mathcal{H}$ is not a polyblock. Then for any finite n, the gap between the approximating polyblock and the feasible set $\mathcal{R} \cap \mathcal{H}$ is nonempty,

$$(\mathcal{P}(\mathcal{V}^{(n)}) \setminus \mathcal{R}) \cap \mathcal{H} \neq \emptyset.$$

As a result, it is always possible to find a continuous increasing function u such that $\overline{u}^{(n)} - \underline{u}^{(n)} > \epsilon$. In other words, without further restrictions, in the worst case the algorithm does not terminate after a finite number of iterations, regardless of the dimension K.

In order to get a meaningful complexity measure, it is obviously necessary to further constrain the class of utility functions. One such possible constraint is to require that the utility function has bounded derivatives. Still, even under this assumption, the use of a black box model immediately implies an exponential lower bound on the complexity of the polyblock algorithm, as shown by the following theorem, which is taken from [20]:

Theorem 3.4.1. *Let $\mathcal{F}(C, P)$ be the class of C-times differentiable functions on $D = [0,1]^K$ whose Cth derivative is bounded by P in the following sense: At any point $\boldsymbol{s} \in \mathcal{D}$ and for any unit vector $\boldsymbol{u}$,*

$$\left| \frac{d^C}{dt^C} f(\boldsymbol{s} + t\boldsymbol{u}) \right| \leq P.$$

Let A be any maximization algorithm that works in the black-box model and yields an ϵ-optimal solution. Then there is a function $f \in \mathcal{F}(C, P)$ such that algorithm A will run for at least

$$B \left(\frac{P}{\epsilon} \right)^{\frac{K}{C}}$$

iterations, where B is a constant that depends on K and C.

Proof. [20, Theorem 4]. □

Instead of maximizing a function f over $\mathcal{D}$, the polyblock algorithm maximizes a function u over wpar $(\mathcal{R}) \cap \mathcal{H}$. According to Proposition 2.3.7, the weak Pareto set is basically a $K - 1$ dimensional manifold. Thus, compared to Theorem 3.4.1, the dimension of the problem is $K - 1$ instead of K. However, this slight reduction in dimensionality does not change the basic result that the polyblock algorithm has at least exponential complexity, even if the derivatives of u are bounded. As a result, computing the global optimum using the polyblock algorithm is practically feasible for small K only. For larger K, it can be expected that the number of iterations required to achieve ϵ-optimality is simply too large, resulting in a run-time of the algorithm that is practically infeasible.

Still, it is important to keep in mind that the above result is a worst-case result. There may well exist nontrivial utility maximization problems $(u, h, \mathcal{R})$ which require much less iterations than the worst case problem.

3.4.6 Related Work

One of the first works that uses monotonic optimization in the context of parameter optimization in wireless systems is [41], where the polyblock algorithm is used in conjunction with a parameterization of the Pareto boundary of the two-user MISO broadcast channel with linear precoding to solve the sum-rate maximization problem. However, results in [41] are limited to the two-user case. In contrast, the framework presented in this chapter covers the general K user case, is applicable to any increasing utility function u, and includes time-sharing/scheduling. The application of the framework to the MISO broadcast channel was first published in [42], and is discussed in detail in Sect. 6.1.

In [43], monotonic optimization is used to solve a utility-based power control problem, given a set of interfering single-input, single-output links. The polyblock algorithm is used to compute the globally optimal power allocation. The algorithm proposed in [43] operates on the $1 + \text{SINR}$ region, hence is very similar to the rate space approach proposed in this chapter. If multiple data streams are decoded in a joint manner (as is usually the case in MIMO systems), SINR usually does not represent a meaningful metric. In this respect, a rate space approach is more general. As discussed in Sect. 4.1.5, using an uplink-downlink duality, a parameterization of the rate region of the MISO broadcast channel in terms of the uplink powers is available, for any number of users K. If time-sharing/scheduling is excluded, by using this parameterization, the utility maximization problem can be written as a power control problem. Hence, the algorithm from [43] can also be used to solve the utility maximization problem in the MISO broadcast channel with linear precoding in case scheduling is not desired. In contrast, [43] is neither applicable to joint scheduling and beamforming nor to utility maximization in the MIMO broadcast channel, nor to any other physical layer configuration in which link rates cannot be expressed as functions of allocated powers.

In [44], monotonic optimization is used to compute the optimal rate allocation in a MIMO broadcast channel with dirty paper coding, in case utilities are nonconcave. A detailed discussion of utility maximization in the MIMO broadcast channel is provided in Sect. 6.2.

A monotonic optimization framework for the MISO *interference channel* (IFC) is proposed in [45]. An achievable rate region based on linear precoding and treating interference as noise is considered. Scheduling is not included. Solutions for the weighted sum-rate problem, a so-called proportional fairness problem,[4] and the max–min problem are provided. Being based on the parameterization of the Pareto boundary of the rate region from [47], the results are again limited to the case of $K = 2$ transmitter/receiver pairs. In order to apply the framework proposed in this section to the MISO IFC model used in [45], a method to solve a generalized max–min problem is required. Such a method is provided in [48], which is a straightforward generalization of the algorithm presented in [49]. The solution method from [48] is applicable to any number of receivers. Moreover, it can easily be generalized to any number of transmitters. The resulting system can be understood as an instance of *network* MISO, where each transmitter corresponds to a base station. From this perspective, the two-user MISO IFC model in [45] is just a special case of a network MISO system. By using the generalized max–min solver from [48], *any* utility maximization problem can be solved using the framework proposed in this section, with *no* restriction on the number of users (other than the

[4] Applying the notion of "proportional fairness" to nonconvex systems is not straightforward, as indicated in Sect. 5.1.2. See [46] for a detailed discussion.

computational complexity of the polyblock algorithm), and, if desired, including scheduling [50].

The monotonic structure that is available in the rate space problem makes monotonic optimization a somewhat natural choice for solving a nonconvex rate space problem to global optimality. Still, other methods from global optimization have been successfully applied to nonconvex utility maximization problems [19]. In [51], a *sum-of-squares* (SOS) method [52] is used to compute upper bounds on the maximally achievable utility in a nonconcave network utility maximization problem. The algorithm is based on solving a sequence of SOS relaxations of increasing degree. For each relaxation of a certain degree, an upper bound on the maximal utility can be computed in polynomial time. The worst-case degree, i.e., the degree needed to compute the global maximum, grows exponentially with the number of links, hence the worst-case complexity of this approach is of the same order as that of the polyblock algorithm. However, numerical results in [51] suggest that the computational effort needed to compute the maximum that is observed in practice is significantly lower than the worst-case bound. The main limitation of the SOS method is the requirement that all problem data be specified in terms of polynomials. With respect to the model used in this monograph, this requires the functions u to be a polynomial and both $h^{-1}(\mathbb{R}_+^K)$ and the rate region $\mathcal{R}$ to be specified by sets of polynomial (in-)equalities. While the restriction to polynomial utility functions may be acceptable, it is generally not possible to specify $\mathcal{R}$ by polynomials. A special case is given by $\mathcal{R}$ being a convex polytope. In this case, $\mathcal{R}$ can be specified by a set of affine inequalities, and the method from [51] becomes applicable. This observation motivates the following generalization of the SOS method to arbitrary convex rate regions: If $\mathcal{R}$ is convex, it can be approximated arbitrarily close by a convex polytope. By iteratively refining the polytope such that the global maximizer of the current polytope approximation is cut away if it is not in $\mathcal{R}$, the global maximizer in $\mathcal{R}$ can be obtained. Clearly, such a method requires repeated application of the SOS method, i.e., in each iteration multiple SOS relaxations have to be solved. How this generalization compares to the polyblock method in terms of computational effort is subject of future research.

Due to the computational complexity of global methods, application of global methods only makes sense in cases where no convex problem formulation can be found. Moreover, the computational complexity of global methods limits their application to off-line computation of performance benchmarks for networks of limited size. In other words, in order to guarantee global optimality in a practical system, problem convexity represents a fundamental requirement. Evidently, for the generic model used in this work, it is not possible to find a convex reformulation for every possible utility maximization problem. For special choices of u, h, and $\boldsymbol{r}$, however, it is possible to find a convex reformulation of a nonconvex parameter space problem.

In the context of utility-based power control for wireless networks, the reformulation approach has been particularly successful, and convex reformulations have been found for a variety of problems [18, 53–55]. For example, in [18, 53] a class of functions u that yields a convex utility maximization problem in the log-power domain is provided. The underlying physical layer model in [18,53] is a network of single antenna nodes, and the physical layer parameters are the transmit powers of the nodes. Each link in the network is assigned a utility value, using a link utility function ψ. The overall system utility is the weighted sum of the link utilities for some given weights, while the link utility function ψ is a function of the SINR on each link. The class of link utility functions that lead to a convex formulation include the so-called α-fair utilities introduced in [56]. Unfortunately, weighted sum rate maximization is not included in the set of problems that have a convex reformulation. Moreover, the results in [18, 53–55] require a specific structure of the rate map $\boldsymbol{r}$—basically, a representation of the physical layer as a network of single-antenna nodes is required, with the transmit power allocation being the only physical layer parameter. This requirement rules out a number of advanced physical layer techniques, such as optimal spatial signal processing in case of multiple antennas at both ends of each link. Still, based on a convex reformulation, [18, 54, 55] can provide distributed algorithms for computing the globally optimal power allocation. In contrast, the global methods discussed in this section are all centralized algorithms of high complexity. Hence, from a practical viewpoint, the methods from [18, 54, 55] are highly relevant, while the application of global methods is limited to the computation of benchmarks (and, possibly, the globally optimal parameter setups can provide a motivation for good heuristics).

In networks with fixed link capacities and nonconcave utilities per flow, a distributed computation of the global optimum is possible for special choices of the overall performance objective. These special cases are utility max–min and utility proportional fairness [7,57,58].

From the discussion of SOS methods, convex reformulations, and utility max–min, it can be observed that in special cases, more efficient solutions than the polyblock algorithm are possible. Still, none of the methods discussed in this section can match the generality of the monotonic optimization approach, as they either make rather restrictive assumptions on the structure of the rate region or the properties of the utility function u. In other words, there clearly exists a trade-off between generality of a solution method and its complexity.

3.5 Local Methods on the Pareto Manifold

Under the assumption that a utility maximization problem $(u, h, \boldsymbol{r}, \mathcal{X}_1)$ exhibits a certain set of properties, the solution methods discussed in Sects. 3.3 and 3.4 provide a parameter setup $\boldsymbol{x}_1^* \in \mathcal{X}_1$ that is globally optimal (up to a tolerance ϵ), i.e.,

$$u(\boldsymbol{r}(\boldsymbol{x}_1^*)) + \epsilon \geq u(\boldsymbol{r}(\boldsymbol{x}_1)), \forall \boldsymbol{x}_1 \in \mathcal{F}_1, \tag{3.86}$$

where $\mathcal{F}_1$ denotes the set of feasible parameter setups, cf. Definition 2.2.1. While the Lagrange duality based method from Sect. 3.3 requires concavity of the functions u and h and convexity of the rate region $\mathcal{R}$, the monotonic optimization framework is applicable to any utility maximization problem, provided there exists an efficient algorithm for solving the intersection problem. As discussed in Sect. 3.4.5, however, the polyblock algorithm is practically feasible for small K only. Moreover, even for small K, the complexity of the algorithm (measured in number of numerical operations until the desired precision is obtained) is rather high. As outlined in Sect. 3.4.5, in the general case of large K and a nonconvex problem setup, seeking a globally optimal solution in the sense of Inequality (3.86) is practically infeasible. Consequently, one has to accept parameter setups as a solution that fulfill a criterion which is significantly weaker than global optimality. An obvious candidate for such a weaker criterion is local optimality, cf. Definition A.7.3.

An obvious approach to finding a parameter setup that is locally optimal is to search for a local maximizer of the composite function $u \circ \boldsymbol{r}$ in $\mathcal{F}_1$. This approach, however, does not take into account the monotonicity structure of the utility maximization problem. As shown by Proposition 2.3.1, the maximum of the rate space problem is attained on the Pareto boundary of the rate region $\mathcal{R}$. In general, the function $u \circ \boldsymbol{r}$ can have any number of local maximizers in $\mathcal{F}_1 \setminus \boldsymbol{r}^{-1}(\text{par}(\mathcal{R}))$. Knowing that the optimal parameter setup is element of $\boldsymbol{r}^{-1}(\text{par}(\mathcal{R}))$, it is reasonable to avoid local maximizers in $\mathcal{F}_1 \setminus \boldsymbol{r}^{-1}(\text{par}(\mathcal{R}))$ as good as possible and to focus on par $(\mathcal{R})$ instead. Based on this observation, instead of searching for a local maximizer of $u \circ \boldsymbol{r}$ in $\mathcal{F}_1$, a method is proposed that aims at finding a local maximizer of u in par $(\mathcal{R}) \cap \mathcal{H}$, with

$$\mathcal{H} = \{\boldsymbol{s} \in \mathcal{D}_h : h(\boldsymbol{s}) \geq 0\}.$$

Thus, again a rate space approach is taken to obtain a solution of the utility maximization problem, and the corresponding rate space problem is given by

$$\max_{\boldsymbol{s}} u(\boldsymbol{s}) \quad \text{s.t.} \quad h(\boldsymbol{s}) \geq 0, \boldsymbol{s} \in \text{par}(\mathcal{R}). \tag{3.87}$$

3.5.1 Optimality Conditions

According to Definition A.7.3, a local optimizer in rate space is a point $\boldsymbol{s}^* \in$ par $(\mathcal{R}) \cap \mathcal{H}$ such that

$$u(\boldsymbol{s}^*) \geq u(\boldsymbol{s}), \forall \boldsymbol{s} \in \mathcal{B}_\eta(\boldsymbol{s}^*) \cap \text{par}(\mathcal{R}) \cap \mathcal{H},$$

for some $\eta > 0$. The definition of a local maximizer is not constructive, in the sense that it does not motivate an algorithm for finding a local maximizer. Instead, first and second order optimality conditions usually provide the foundation for an algorithmic solution, under the assumption that first and second order information about the involved mathematical objects is available. Considering Problem (3.87), these objects are the functions u and h and the Pareto set par $(\mathcal{R})$.

In the following, a number of assumptions on the properties of these objects are made, which finally yields a problem formulation that is amenable to an algorithmic solution. Throughout this section, let $\mathcal{T}_{\mathcal{S}}(\boldsymbol{s})$ denote the tangent cone of $\mathcal{S}$ at $\boldsymbol{s}$, cf. Definition A.7.4.

Assumption 3.5.1. *The utility function u is continuously differentiable.*

The following first order necessary optimality condition is obtained:

Proposition 3.5.1. *Let $\boldsymbol{s}$ be a local maximizer of u in* $\operatorname{par}(\mathcal{R}) \cap \mathcal{H}$. *Then*

$$\nabla u(\boldsymbol{s})^{\mathrm{T}} \boldsymbol{v} \leq \mathbf{0}, \forall \boldsymbol{v} \in \mathcal{T}_{\operatorname{par}(\mathcal{R}) \cap \mathcal{H}}(\boldsymbol{s}). \tag{3.88}$$

Proof. [25, Theorem 5.1.2]. □

According to Definition A.7.5, a point $\boldsymbol{s} \in \operatorname{par}(\mathcal{R}) \cap \mathcal{H}$ satisfying (3.88) is a stationary point in $\operatorname{par}(\mathcal{R}) \cap \mathcal{H}$. By definition of a utility maximization problem, there always exists a global maximizer. As a consequence, the set of stationary points is nonempty. Stationarity is an even weaker notion than local optimality: Clearly, every local maximizer is a stationary point, but not every stationary point is a local maximizer. Still, in the following, a stationary point is accepted as local solution of a utility maximization problem:

Definition 3.5.1 (Local Solution). Let $\boldsymbol{x}_1^* \in \mathcal{X}_1$ be a parameter setup such that $\boldsymbol{r}(\boldsymbol{x}_1^*)$ is a stationary point of u in $\operatorname{par}(\mathcal{R}) \cap \mathcal{H}$. Then $\boldsymbol{x}_1^*$ is a *local solution* of the utility maximization problem $(u, h, \boldsymbol{r}, \mathcal{X}_1)$.

The choice of stationarity as the solution criterion is based on the observation that a stationary point is defined by the properties of the tangent cone of $\operatorname{par}(\mathcal{R}) \cap \mathcal{H}$ and the gradient of u. As shown in the following, under suitable differentiability assumptions, the tangent cone can be completely characterized by first-order information. In contrast, verifying local optimality generally requires at least second-order information [25]. Consequently, stationarity can be achieved based on a minimal set of assumptions, while guaranteeing local optimality necessitates further assumptions on the properties of the utility maximization problems under consideration.

In (3.88), the tangent cone of $\operatorname{par}(\mathcal{R}) \cap \mathcal{H}$ at $\boldsymbol{s}$ is still an abstract object. In the next step, a number of assumptions are made in order to equip the tangent cone $\mathcal{T}_{\operatorname{par}(\mathcal{R}) \cap \mathcal{H}}(\boldsymbol{s})$ with the necessary structure such that a stationary point in $\operatorname{par}(\mathcal{R}) \cap \mathcal{H}$ can be computed by a numerical algorithm. By Corollary A.7.1,

$$\mathcal{T}_{\operatorname{par}(\mathcal{R}) \cap \mathcal{H}}(\boldsymbol{s}) \subseteq \mathcal{T}_{\operatorname{par}(\mathcal{R})}(\boldsymbol{s}) \cap \mathcal{T}_{\mathcal{H}}(\boldsymbol{s}). \tag{3.89}$$

The following assumption provides a basis for a convenient representation of $\mathcal{T}_{\mathcal{H}}(\boldsymbol{s})$.

Assumption 3.5.2. *The function h is the point-wise minimum over the component functions of an increasing, continuously differentiable function $\boldsymbol{h} : \mathcal{D}_{\boldsymbol{h}} \rightarrow \mathbb{R}^M$, i.e.,*

$$h = \min_{m \in \{1,\dots,M\}} h_m(\cdot),$$

for some $M \geq 1$, and $\mathcal{D}_h \supset \mathbb{R}_+^K$ open.

Assumption 3.5.2 implies that

$$\mathcal{H} = \bigcap_{m=1}^{M} \{s \in \mathcal{D}_h : h_m(s) \geq 0\} = \bigcap_{m=1}^{M} h_m^{-1}(\mathbb{R}_+).$$

According to Corollary A.7.1,

$$\mathcal{T}_{\mathcal{H}}(s) \subseteq \bigcap_{m=1}^{M} \mathcal{T}_{h_m^{-1}(\mathbb{R}_+)}(s). \tag{3.90}$$

For convenience, define

$$\mathcal{H}'(s) = \bigcap_{m=1}^{M} \mathcal{T}_{h_m^{-1}(\mathbb{R}_+)}(s).$$

From differentiability of $\boldsymbol{h}$ it follows that

$$\mathcal{T}_{h_m^{-1}(\mathbb{R}_+)}(s) = \begin{cases} \left\{v \in \mathbb{R}^K : \nabla h_m(s)^{\mathrm{T}} v \geq 0\right\}, & h_m(s) = 0, \\ \mathbb{R}^K, & h_m(s) > 0. \end{cases}$$

Let the set $\mathcal{J}(s)$ contain the indices of the constraints $h_m(s) \geq 0$ that are active at $s \in \mathbb{R}_+^K$, i.e.,

$$\mathcal{J}(s) = \{m : h_m(s) = 0\} \subseteq \{1, \dots, M\}.$$

Then

$$\mathcal{H}'(s) = \bigcap_{m \in \mathcal{J}(s)} \left\{v \in \mathbb{R}^K : \nabla h_m(s)^{\mathrm{T}} v \geq 0\right\}. \tag{3.91}$$

Combining (3.89) and (3.90) yields

$$\mathcal{T}_{\mathrm{par}(\mathcal{R}) \cap \mathcal{H}}(s) \subseteq \mathcal{T}_{\mathrm{par}(\mathcal{R})}(s) \cap \mathcal{H}'(s). \tag{3.92}$$

In the following, a representation of $\mathcal{T}_{\mathrm{par}(\mathcal{R}) \cap \mathcal{H}}(s)$ as the intersection of a linear space with a set of half-spaces is derived. The derivation is based on (3.92) and proceeds in two steps. According to 3.91, $\mathcal{H}'(s)$ has a representation in terms of an intersection of half-spaces. Accordingly, in the first step, par$(\mathcal{R})$ is equipped with some additional structure such that $\mathcal{T}_{\text{par}(\mathcal{R})}(\boldsymbol{s})$ is a linear space. In the second step, a sufficient condition for equality to hold in (3.92) is provided.

Assumption 3.5.3. *The weak Pareto set and the Pareto set are identical,*

$$\text{wpar}(\mathcal{R}) = \text{par}(\mathcal{R}).$$

In Sect. 2.3, it is shown that the intersection of the weak Pareto set of a proper set and $\mathbb{R}^K_{++}$ is a $K-1$ dimensional manifold. Assumption 3.5.3 provides the same property for the Pareto set. In the following, let $\mathcal{M}$ denote the Pareto manifold:

$$\mathcal{M} = \text{par}(\mathcal{R}) \cap \mathbb{R}^K_{++}.$$

According to (3.91), the differentiability assumption on $\boldsymbol{h}$ provides a representation of $\mathcal{H}'(\boldsymbol{s})$ as an intersection of half-spaces. Likewise, in order to obtain a convenient representation of the tangent cone of par$(\mathcal{R})$, the Pareto manifold $\mathcal{M}$ is equipped with a differentiable structure.

Assumption 3.5.4. *The Pareto manifold $\mathcal{M}$ is a C^1 submanifold of $\mathbb{R}^K$.*

Proposition 3.5.2. *There exists a continuous map $\boldsymbol{n}: \mathcal{M} \to \mathbb{R}^K_+$ such that*

$$\|\boldsymbol{n}(\boldsymbol{s})\|_2 = 1, \tag{3.93}$$

$$\mathcal{T}_{\mathcal{M}}(\boldsymbol{s}) = \text{null}\left(\boldsymbol{n}(\boldsymbol{s})^{\text{T}}\right). \tag{3.94}$$

Proof. By Assumption 3.5.4, $\mathcal{M}$ is a $K-1$-dimensional C^1 submanifold of $\mathbb{R}^K$, hence, by Proposition A.8.4 there exists a continuous map $\boldsymbol{n}: \mathcal{M} \to \mathbb{R}^K$ satisfying (3.93) and (3.94). It remains to show that there exists a continuous map with $\boldsymbol{n}(\boldsymbol{s}) \in \mathbb{R}^K_+, \forall \boldsymbol{s} \in \mathcal{M}$. According to Proposition A.7.2, $\mathcal{T}_{\mathcal{M}}(\boldsymbol{s})$ separates $\mathbb{R}^K_+$ and $-\mathbb{R}^K_+$. As a consequence, $\boldsymbol{n}(\boldsymbol{s})$ has to fulfill either $\boldsymbol{n}(\boldsymbol{s}) \in \mathbb{R}^K_+$ or $\boldsymbol{n}(\boldsymbol{s}) \in -\mathbb{R}^K_+$. The sets

$$\mathcal{N}_+ = \left\{\boldsymbol{n} \in \mathbb{R}^K_+ : \|\boldsymbol{n}\|_2 = 1\right\}$$

and

$$\mathcal{N}_- = \left\{\boldsymbol{n} \in -\mathbb{R}^K_+ : \|\boldsymbol{n}\|_2 = 1\right\}$$

are not connected. Hence, by continuity of $\boldsymbol{n}$, either $\boldsymbol{n}(\mathcal{M}) \subset \mathbb{R}^K_+$ or $\boldsymbol{n}(\mathcal{M}) \subset -\mathbb{R}^K_+$. Assume $\boldsymbol{n}(\mathcal{M}) \subset -\mathbb{R}^K_+$. Then the map $-\boldsymbol{n}$ is continuous and $-\boldsymbol{n}(\mathcal{M}) \subset \mathbb{R}^K_+$. □

Example 3.5.1. Proposition 3.5.2 is illustrated in Fig. 3.3. Figure 3.3a shows a proper rate region for which $\mathcal{M} = \text{par}(\mathcal{R}) \cap \mathbb{R}^2_{++}$ is a 1-dimensional C^1 submanifold of $\mathbb{R}^2$. Also shown are the normal vectors $\boldsymbol{n}(\boldsymbol{s})$ at some points $\boldsymbol{s}$ in $\mathcal{M}$. It can be observed that the normal vectors vary continuously on $\mathcal{M}$. Moreover, all normal

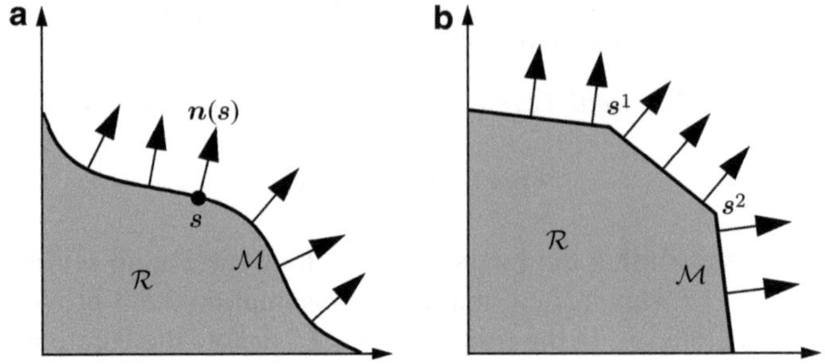

Fig. 3.3 Two proper rate regions with their Pareto sets. In (**a**), the Pareto manifold $\mathcal{M}$ is a C^1 submanifold of $\mathbb{R}^2$, while in (**b**), the Pareto manifold $\mathcal{M}$ is not a C^1 submanifold

vectors point to the upper right, as they are elements of $\mathbb{R}^2_+$. Finally, at each $\boldsymbol{s} \in \mathcal{M}$, the tangent cone is clearly a line orthogonal to the normal vector.

Figure 3.3b shows a proper rate region for which $\mathcal{M}$ is *not* a C^1 submanifold of $\mathbb{R}^2$. As a result, it is not possible to find a continuous normal field, as can be seen around $\boldsymbol{s}^1$ and $\boldsymbol{s}^2$. Moreover, at the points $\boldsymbol{s}^1$ and $\boldsymbol{s}^2$, the tangent cone is not a 1-dimensional subspace, hence cannot be characterized as the null space of a normal vector.

Corollary 3.5.3. *Let* $\boldsymbol{n} : \mathcal{M} \to \mathbb{R}^K_+$ *be a continuous map satisfying* (3.93) *and* (3.94). *Then* $\boldsymbol{n}$ *is the only such map.*

Definition 3.5.2. Let $\boldsymbol{n} : \mathcal{M} \to \mathbb{R}^K_+$ be the unique continuous map satisfying (3.93) and (3.94). The map $\boldsymbol{n}$ is the *field of normal vectors* of $\mathcal{M}$.

As shown by Proposition 3.5.2, Assumption 3.5.4 implies that for all $\boldsymbol{s} \in \mathcal{M}$, the tangent cone of par $(\mathcal{R})$ at $\boldsymbol{s}$ can be represented as

$$\mathcal{T}_{\text{par}(\mathcal{R})}(\boldsymbol{s}) = \left\{\boldsymbol{v} \in \mathbb{R}^K : \boldsymbol{n}(\boldsymbol{s})^{\mathrm{T}}\boldsymbol{v} = 0\right\},$$

which finally is the desired representation of $\mathcal{T}_{\text{par}(\mathcal{R})}(\boldsymbol{s})$. This representation, however, is only valid for $\boldsymbol{s} \in \mathcal{M} \subset \mathbb{R}^K_{++}$. Under the assumption that the set $\mathcal{H}$ is a subset of $\mathbb{R}^K_{++}$, it is sufficient to have a representation of $\mathcal{T}_{\text{par}(\mathcal{R})}(\boldsymbol{s})$ for strictly positive $\boldsymbol{s}$ only. In the general case, however, solutions $\boldsymbol{s} \notin \mathbb{R}^K_{++}$ cannot be excluded. Accordingly, a two-step approach is taken: First, a method is developed under the assumption $\mathcal{H} \in \mathbb{R}^K_{++}$. In a second step, a heuristic is introduced such that the proposed method can converge to points on the boundary of $\mathbb{R}^K_+$, cf. Sect. 3.5.5.

Assumption 3.5.5. *All feasible points are strictly positive, i.e.,*

$$\mathcal{H} \in \mathbb{R}^K_{++}.$$

Given a set $\mathcal{H}$ which does not fulfill the strict positivity assumption, the assumption can be fulfilled by adding additional constraints

$$s - \eta \mathbf{1} \geq \mathbf{0},$$

with $\eta > 0$, yielding a new problem with

$$\tilde{\mathcal{H}} = \mathcal{H} \cap \mathcal{A} \subset \mathbb{R}^K_{++},$$

where

$$\mathcal{A} = \{\eta \mathbf{1}\} + \mathbb{R}^K_{+}.$$

In order to obtain a well-defined utility maximization problem, $\mathcal{R} \cap \tilde{\mathcal{H}} \neq \emptyset$ must hold true, i.e., η must be sufficiently small. Clearly, adding these constraints excludes any (existing) stationary points of the original problem in $\mathcal{H} \setminus \mathcal{A}$, and may add stationary points on the boundary of $\tilde{\mathcal{H}}$. This issue is addressed in Sect. 3.5.5.

Under Assumption 3.5.5, the rate space problem can be written as

$$\max_{s} u(s) \quad \text{s.t.} \quad h(s) \geq \mathbf{0}, s \in \mathcal{M}. \tag{3.95}$$

The set of feasible rates in Problem (3.95) is given by the intersection of the Pareto manifold $\mathcal{M}$ and the preimage of $\mathbb{R}^M_+$ under a continuously differentiable constraint function h. If the manifold constraint was absent, Problem (3.95) would be a standard nonlinear program [25]: maximize a differentiable function under nonlinear inequality constraints specified by a differentiable constraint function, with optimization variable $x \in \mathbb{R}^K$. A stationary point of such a problem can be computed using a standard algorithm, usually by computing a Karush-Kuhn-Tucker (KKT) point and verifying a constraint qualification [25]. The manifold constraint makes the rate space problem non-standard. A constraint that is not expressed by a constraint function, but by membership in an abstract set $\mathcal{X}$ is frequently denoted as an "abstract" constraint [28]. By Assumption 3.5.4, the abstract constraint is equipped with a differentiable structure. In the following, it is shown that by exploiting this structure, it is possible to solve the rate space problem with the same methods as a standard nonlinear program.

In the following, let $n : \mathcal{M} \to \mathbb{R}^K_+$ be the field of normal vectors of $\mathcal{M}$.

Proposition 3.5.4. *Let $s \in \mathcal{M} \cap \mathcal{H}$. Then*

$$\nabla u(s)^{\mathrm{T}} v \leq 0, \forall v \in \mathcal{T}_{\mathcal{M}}(s) \cap \mathcal{H}'(s). \tag{3.96}$$

if and only if there exists $\lambda \in \mathbb{R}^M_+$ and $\nu \in \mathbb{R}$ such that

$$\nabla u(s) + \sum_{m \in \mathcal{J}(s)} \lambda_m \nabla h_m(s) + \nu n(s) = \mathbf{0}. \tag{3.97}$$

Proof. Farkas' Theorem see, e.g., [25]. □

Remark 3.5.1. Proposition 3.5.4 is very similar to the standard KKT conditions for the maximization of a differentiable function under equality and inequality constraints with differentiable constraint functions. This similarity is not surprising. In fact, assume that $\mathcal{M}$ is the solution set of an equality constraint with differentiable constraint function $c : \mathbb{R}^K \to \mathbb{R}$:

$$\mathcal{M} = \left\{ \boldsymbol{s} \in \mathbb{R}^K : c(\boldsymbol{s}) = 0 \right\} = c^{-1}(\{0\}).$$

In addition, assume that

$$\nabla c(\boldsymbol{s}) \neq \mathbf{0}, \forall \boldsymbol{s} \in \mathcal{M}.$$

Then $\mathcal{M}$ is a $K-1$ dimensional differentiable manifold [59], and the tangent space of $\mathcal{M}$ at $\boldsymbol{s} \in \mathcal{M}$ is given by

$$\mathcal{T}_{\mathcal{M}}(\boldsymbol{s}) = \left\{ \boldsymbol{v} \in \mathbb{R}^K : \nabla c(\boldsymbol{s})^{\mathrm{T}} \boldsymbol{v} = 0 \right\}.$$

Similarly, in [60], the KKT conditions for problems with differentiable constraint function $\boldsymbol{c} : \mathbb{R}^K \to \mathbb{R}^M$ are developed based on recognizing the set $\boldsymbol{c}^{-1}(\{\mathbf{0}\})$ as a differentiable manifold.

For the derivation of the KKT conditions it does not play a role whether the manifold is the solution set of an equality constraint or the Pareto set of a proper rate region, as long as the required objects to describe the tangent space are available. In case that the manifold is of dimension $K-1$, the tangent space at $\boldsymbol{s}$ can be represented by a normal vector. If the manifold is generated by an equality constraint, such a normal vector is given by the gradient of the constraint function at $\boldsymbol{s}$,

$$\boldsymbol{n}(\boldsymbol{s}) = \frac{\nabla c(\boldsymbol{s})}{\|\nabla c(\boldsymbol{s})\|_2}, \tag{3.98}$$

Accordingly, plugging (3.98) into (3.97) yields the standard KKT conditions for a problem with a single equality constraint $c(\boldsymbol{s}) = 0$.

Definition 3.5.3. A point $\boldsymbol{s} \in \mathcal{M}$ satisfying (3.96) is denoted as a *KKT point of the rate space problem.*

As a consequence of (3.92), every KKT point is a stationary point, while a stationary point is not necessarily a KKT point. Thus, while there always exists a stationary point by definition of a utility maximization problem, the set of KKT points may be empty. On the other hand, the KKT conditions are the foundation for an algorithmic solution. As usual, to make searching for a KKT point in $\mathcal{M}$ a meaningful endeavor, a constraint qualification is required. Based on the characterization of $\mathcal{T}_{\mathcal{M}}(\boldsymbol{s})$ as the nullspace of $\boldsymbol{n}(\boldsymbol{s})^{\mathrm{T}}$, a sufficient condition for equality to hold in (3.92) is provided by the following proposition:

Proposition 3.5.5. *Let $\boldsymbol{s}$ be a point in $\mathcal{M} \cap \mathcal{H}$. If the vectors in the set*

$$\{\nabla h_m(\boldsymbol{s})\}_{m\in\mathcal{J}(\boldsymbol{s})} \cup \{\boldsymbol{n}(\boldsymbol{s})\}$$

are linearly independent, then

$$\mathcal{T}_{\mathcal{M}\cap\mathcal{H}}(\boldsymbol{s}) = \mathcal{T}_{\mathcal{M}}(\boldsymbol{s}) \cap \mathcal{H}'(\boldsymbol{s}). \tag{3.99}$$

Proof. From $\mathcal{M} \cap \mathcal{H} \subset \mathbb{R}^K_{++}$ it follows that there exists an open set $\mathcal{X} \subset \mathbb{R}^K_+$ such that $\mathcal{M} \cap \mathcal{H} \subset \mathcal{X}$. The function $\boldsymbol{h}$ is continuous, implying that h_m is continuous for all $m \notin \mathcal{J}(\boldsymbol{s})$. Moreover, by Proposition 3.5.2, $\boldsymbol{n}$ is continuous. As a result, the *Linear Independence Constraint Qualification* [25] holds true, which implies (3.99). □

3.5.2 Computing a KKT Point on the Pareto Manifold

In order to find a KKT point on $\mathcal{M}$, a means for moving on $\mathcal{M}$ is required. By assumption, $\mathcal{M}$ is a $K-1$-dimensional C^1 submanifold of $\mathbb{R}^K$. Hence, by Proposition A.8.2, at every point $\boldsymbol{s} \in \mathcal{M}$, there exists a local parameterization $(\mathcal{U}, \boldsymbol{\psi})$ such that $\boldsymbol{s} \in \boldsymbol{\psi}(\mathcal{U})$ and $\boldsymbol{\psi}$ is of class C^1. By varying the local coordinates $\boldsymbol{\mu} \in \mathcal{U} \subseteq \mathbb{R}^{K-1}$, one can move on $\mathcal{M}$, as

$$\boldsymbol{\psi}(\mathcal{U}) \subseteq \mathcal{M}.$$

According to Proposition A.8.3, the tangent space of $\mathcal{M}$ at $\boldsymbol{\psi}(\boldsymbol{\mu})$ is given by the range of the Jacobian of $\boldsymbol{\psi}$ at $\boldsymbol{\mu}$,

$$\mathcal{T}_{\mathcal{M}}(\boldsymbol{\psi}(\boldsymbol{\mu})) = \operatorname{range}\left(\nabla\boldsymbol{\psi}^{\mathrm{T}}(\boldsymbol{\mu})\right). \tag{3.100}$$

Equations (3.94) and (3.100) imply

$$\operatorname{range}\left(\nabla\boldsymbol{\psi}_{\boldsymbol{s}}^{\mathrm{T}}(\boldsymbol{\mu})\right) = \operatorname{null}\left(\boldsymbol{n}(\boldsymbol{\psi}_{\boldsymbol{s}}(\boldsymbol{\mu}))^{\mathrm{T}}\right). \tag{3.101}$$

In general, $\boldsymbol{\psi}(\mathcal{U})$ is a strict subset of $\mathcal{M}$, i.e., a single parameterization is not sufficient to cover $\mathcal{M}$. However, Proposition 2.3.7 shows that for the manifold $\mathcal{M}$ under consideration, there exists a parameterization that completely covers $\mathcal{M}$. As demonstrated by Proposition 2.3.7, the existence of such global parameterizations stems from the fact that the underlying rate region is proper.

Assume that a global parameterization $(\mathcal{U}, \boldsymbol{\psi})$ is given. Moreover, assume that $\boldsymbol{\psi}$ is of class C^1, which implies that $\nabla\boldsymbol{\psi}(\boldsymbol{\mu})$ is well-defined and of rank $K-1$ for all

$\boldsymbol{\mu} \in \mathcal{U}$. A general method to construct such a *differentiable global parameterization* is provided in Sect. 3.5.3. Note that by Assumptions 3.5.5 and 3.5.3,

$$\mathcal{H} \cap \text{par}(\mathcal{R}) = \mathcal{H} \cap \mathcal{M}.$$

Thus,

$$\boldsymbol{\psi}(\mathcal{U}) \supseteq \mathcal{H} \cap \mathcal{M},$$

i.e., the parameterization $\boldsymbol{\psi}$ covers all feasible points of the rate space problem (3.87). Plugging $\boldsymbol{\psi}$ into the rate space problem (3.95) yields the optimization problem

$$\max_{\boldsymbol{\mu} \in \mathbb{R}^{K-1}} u(\boldsymbol{\psi}(\boldsymbol{\mu})) \quad \text{s.t.} \quad \boldsymbol{h}(\boldsymbol{\psi}(\boldsymbol{\mu})) \geq \mathbf{0}. \tag{3.102}$$

By using a global parameterization, the manifold $\mathcal{M}$ can be completely eliminated from the optimization problem. As a result, Problem (3.102) is a standard maximization problem with inequality constraints in $\mathbb{R}^{K-1}$, with differentiable cost function $f = u \circ \boldsymbol{\psi}$ and differentiable constraint function $\boldsymbol{c} = \boldsymbol{h} \circ \boldsymbol{\psi}$. According to the chain rule, the gradients of f and $\boldsymbol{c}$ are given by

$$\nabla f(\boldsymbol{\mu}) = \nabla \boldsymbol{\psi}(\boldsymbol{\mu}) \nabla u(\boldsymbol{\psi}(\boldsymbol{\mu})),$$
$$\nabla \boldsymbol{c}(\boldsymbol{\mu}) = \nabla \boldsymbol{\psi}(\boldsymbol{\mu}) \nabla \boldsymbol{h}(\boldsymbol{\psi}(\boldsymbol{\mu})).$$

Proposition 3.5.6. *A point $\boldsymbol{\mu} \in \mathcal{U}$ is a KKT point of Problem* (3.102) *if and only if $\boldsymbol{\psi}(\boldsymbol{\mu})$ is a KKT point of the rate space problem.*

Proof. A KKT point of Problem (3.102) is a point $\boldsymbol{\mu} \in \mathbb{R}^{K-1}$ such that $\boldsymbol{c}(\boldsymbol{\mu}) \geq \mathbf{0}$ and there exists $\boldsymbol{\lambda} \geq \mathbf{0}$ such that

$$\nabla f(\boldsymbol{\mu}) + \nabla \boldsymbol{c}(\boldsymbol{\mu})\boldsymbol{\lambda} = \mathbf{0}, \tag{3.103}$$
$$\lambda_m c_m(\boldsymbol{s}) = 0, \forall m = 1, \ldots, M.$$

Define

$$\boldsymbol{L}(\boldsymbol{s}, \boldsymbol{\lambda}, \nu) = \nabla u(\boldsymbol{s}) + \nabla \boldsymbol{h}(\boldsymbol{s})\boldsymbol{\lambda} + \nu \boldsymbol{n}(\boldsymbol{s}).$$

Recall from (3.101) that

$$\nabla \boldsymbol{\psi}(\boldsymbol{\mu}) \boldsymbol{n}(\boldsymbol{\psi}(\boldsymbol{\mu})) = \mathbf{0}.$$

First, assume that $\boldsymbol{\psi}(\boldsymbol{\mu})$ is a KKT point of the rate space problem. By Proposition 3.5.4, there exists $\boldsymbol{\lambda} \in \mathbb{R}_+^K$ and $\nu \in \mathbb{R}$ such that

$$L(\psi(\mu), \lambda, \nu) = \mathbf{0},$$
$$\lambda_m h_m(\psi(\mu)) = 0, \forall m = 1, \ldots, M.$$

Then

$$\nabla\psi(\mu) L(\psi(\mu), \lambda, \nu) = \nabla f(\mu) + \nabla c(\mu)\lambda = \mathbf{0},$$

implying that μ is a KKT point of Problem (3.102).

Second, assume that μ is a KKT point of Problem (3.102). Let

$$W = \left[\nabla\psi(\mu)^{\mathrm{T}} \; n(\psi(\mu))\right].$$

Then

$$W^{\mathrm{T}} L(\psi(\mu), \lambda, \nu) = \begin{bmatrix} \nabla f(\mu) + \nabla c(\mu)\lambda \\ n(\psi(\mu))^{\mathrm{T}} \nabla u(\psi(\mu)) + n(\psi(\mu))^{\mathrm{T}} \nabla h(\psi(\mu))\lambda + \nu \end{bmatrix}.$$

Using (3.103) and setting

$$\nu = -n(\psi(\mu))^{\mathrm{T}} \nabla u(\psi(\mu)) - n(\psi(\mu))^{\mathrm{T}} \nabla h(\psi(\mu))\lambda$$

yields

$$W^{\mathrm{T}} L(\psi(\mu), \lambda, \nu) = \mathbf{0}.$$

The matrix W is invertible, which implies

$$L(\psi(\mu), \lambda, \nu) = \mathbf{0}.$$

As a result, by Proposition 3.5.4, $\psi(\mu)$ is a KKT point of the rate space problem.

□

Based on Proposition 3.5.6, a KKT point of the rate space problem can be obtained by using a standard algorithm for computing a KKT point of Problem (3.102). Any algorithm that meets the following requirements can be employed:

(1) The algorithm needs only first-order information. This constraint stems from the fact that the parameterizations considered in this work are of class C^1, an immediate result of the assumption that $\mathcal{M}$ is a C^1 submanifold of $\mathbb{R}^K$. Of course, higher-order information can be made available by assuming that $\mathcal{M}$ is a C^r manifold of $\mathbb{R}^K$, with $r > 1$. However, in this work only the minimal requirement C^1 is considered.
(2) The algorithm generates a sequence of feasible points. Infeasible points may lie outside the coverage of the parameterization, therefore feasibility of iterates represents a basic requirement for the parameterization-based approach.

One example for an algorithm that meets above requirements is Rosen's Gradient Projection Method [61].

To summarize, based on Assumptions 3.5.2–3.5.5 and the assumption that a differentiable global parameterization is available, the non-standard rate space problem (3.87) can be transformed into a standard inequality constrained optimization problem, and a KKT point can be computed by using standard algorithms. The benefit of transforming the original problem into a standard problem becomes evident when considering the difficulties in proving the theoretical convergence properties of some standard algorithms for nonlinear programming, cf. [62–65]. By transforming the rate space problem into a standard problem, a re-use of already established results is possible.

The key element in going from the search for a KKT point on a differentiable manifold to a standard inequality-constrained optimization problem is the differentiable global parameterization $(\mathcal{U}, \boldsymbol{\psi})$. In the following, a generic method for constructing such a parameterization is presented.

3.5.3 A Generic Global Parameterization

As discussed in Sect. 3.5.2, differentiable global paramterizations enable the use of standard algorithms for the computation of a KKT point in $\mathcal{M}$. This section presents a generic method for constructing a differentiable global parameterization. The method is generic in the sense that it can be applied to any $K-1$ dimensional C^1 submanifold of $\mathbb{R}^K$ that is compatible with the structural assumptions from Sect. 3.5.1.

The outline of this section is as follows: First, an open subset $\mathcal{U}$ of $\mathbb{R}^{K-1}$ and a map $\boldsymbol{\psi}_{\boldsymbol{s}}$ from $\mathcal{U}$ to $\mathcal{M}$ are defined. Then it is shown that the pair $(\mathcal{U}, \boldsymbol{\psi}_{\boldsymbol{s}})$ is a differentiable global parameterization of $\mathcal{M}$. Finally, a method for computing the gradient matrix $\nabla\boldsymbol{\psi}_{\boldsymbol{s}}(\boldsymbol{\mu})$ is provided.

For $\boldsymbol{s} \in \mathcal{M}$, let the columns of $\boldsymbol{Q}_{\boldsymbol{s}} \in \mathbb{R}^{K \times K-1}$ constitute an orthonormal basis of null $\left(\boldsymbol{n}(\boldsymbol{s})^{\mathrm{T}}\right)$. Pick any $\boldsymbol{s} \in \mathcal{M}$ and define a function $\boldsymbol{g} : \mathbb{R}^{K-1} \to \mathbb{R}^K_+$ as follows:

$$\boldsymbol{g}(\boldsymbol{\mu}) = \boldsymbol{\phi}_{\boldsymbol{\mu}}(\delta(\boldsymbol{\mu})),$$

where $\boldsymbol{\phi}_{\boldsymbol{\mu}} : \mathbb{R} \to \mathbb{R}^K_+$,

$$\boldsymbol{\phi}_{\boldsymbol{\mu}}(\gamma) = (\boldsymbol{s} + \boldsymbol{Q}_{\boldsymbol{s}}\boldsymbol{\mu} + \boldsymbol{n}(\boldsymbol{s})\gamma)^+ ,$$

and $\delta : \mathbb{R}^{K-1} \to \mathbb{R}$,

$$\delta(\boldsymbol{\mu}) = \max\left\{\gamma \in \mathbb{R} : \boldsymbol{\phi}_{\boldsymbol{\mu}}(\gamma) \in \mathcal{R}\right\}. \tag{3.104}$$

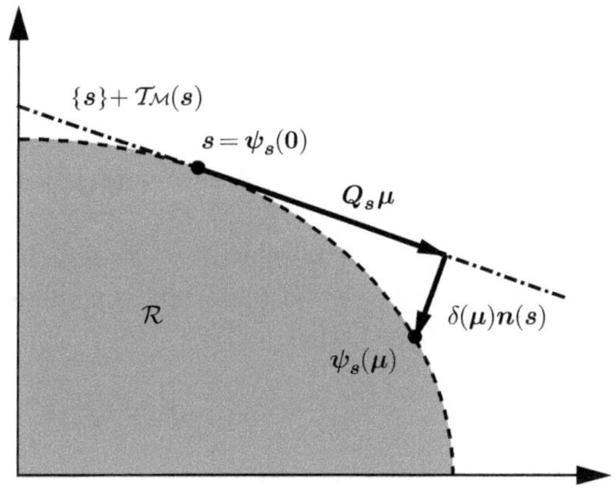

Fig. 3.4 The map ψ_s: Starting at $s \in \mathcal{M}$, a step $Q_s\mu$ in a tangent direction is followed by a corrector step $\delta(\mu)n(s)$ such that $\psi(\mu) \in \mathcal{M}$

According to Proposition A.4.4, $g(\mu) \in \text{wpar}(\mathcal{R})$, which under Assumption 3.5.3 implies $g(\mu) \in \text{par}(\mathcal{R})$. Define

$$\mathcal{U} = g^{-1}(\mathcal{M}),$$

and $\psi_s : \mathcal{U} \to \mathcal{M}$,

$$\psi_s(\mu) = s + Q_s\mu + n(s)\delta(\mu). \tag{3.105}$$

Note that $\psi_s(\mu) = g(\mu), \forall \mu \in \mathcal{U}$, due to the fact that

$$s + Q_s\mu + n(s)\delta(\mu) \in \mathbb{R}^K_{++}, \forall \mu \in \mathcal{U}.$$

The map ψ_s is illustrated in Fig. 3.4.

Lemma 3.5.7. *The map ψ_s is a bijection. The inverse map $\psi_s^{-1} : \mathcal{M} \to \mathcal{U}$ is given by*

$$\psi_s^{-1}(s') = Q_s^{\mathrm{T}}(s' - s). \tag{3.106}$$

Proof. The map ψ_s is a bijection if and only if the preimage of s' is a singleton for all $s' \in \mathcal{M}$. Let $s' \in \mathcal{M}$. The equation

$$s' = s + Q_s\mu + n(s)\gamma$$

has a unique solution (μ', γ'), due to the fact that the columns of Q_s and $n(s)$ constitute a basis of $\mathbb{R}^K$. Thus,

$$\psi_s^{-1}(\{s'\}) \subseteq \{\mu'\}.$$

In order to have $\psi_{\boldsymbol{s}}^{-1}(\{\boldsymbol{s}'\}) = \{\boldsymbol{\mu}'\}$, the equality $\gamma' = \delta(\boldsymbol{\mu}')$ is required. By definition of the function δ, $\gamma' \leq \delta(\boldsymbol{\mu}')$. Now assume $\gamma' < \delta(\boldsymbol{\mu}')$. This implies that there exists $\tilde{\boldsymbol{s}} \in \mathcal{R}$ such that

$$\tilde{\boldsymbol{s}} = \boldsymbol{s}' + (\delta(\boldsymbol{\mu}') - \gamma')\boldsymbol{n}(\boldsymbol{s}).$$

But $\boldsymbol{n}(\boldsymbol{s}) \in \mathbb{R}_+^K \setminus \boldsymbol{0}$ implies $\tilde{\boldsymbol{s}} > \boldsymbol{s}'$, which contradicts $\boldsymbol{s}' \in \operatorname{par}(\mathcal{R})$. Thus, the preimage of any $\boldsymbol{s}' \in \mathcal{M}$ under $\psi_{\boldsymbol{s}}$ is a singleton. Due to the orthogonality of the columns of $\boldsymbol{Q}_{\boldsymbol{s}}$ and $\boldsymbol{n}(\boldsymbol{s})$,

$$\boldsymbol{\mu}' = \boldsymbol{Q}_{\boldsymbol{s}}^{\mathrm{T}}(\boldsymbol{s}' - \boldsymbol{s}).$$

□

Lemma 3.5.8. *Let* $\boldsymbol{A}, \boldsymbol{B}_{\boldsymbol{s}'} \in \mathbb{R}^{K \times K-1}$, $\mathbf{c}, \boldsymbol{d} \in \mathbb{R}^K \setminus \{\boldsymbol{0}\}$ *such that*

$$\operatorname{rank}(\boldsymbol{A}) = \operatorname{rank}(\boldsymbol{B}_{\boldsymbol{s}'}) = K - 1,$$
$$\mathbf{c}^{\mathrm{T}}\boldsymbol{A} = \boldsymbol{d}^{\mathrm{T}}\boldsymbol{B}_{\boldsymbol{s}'} = \boldsymbol{0}.$$

Then $\operatorname{rank}\left(\boldsymbol{A}^{\mathrm{T}}\boldsymbol{B}_{\boldsymbol{s}'}\right) = K - 1$ *if and only if* $\mathbf{c}^{\mathrm{T}}\boldsymbol{d} \neq 0$.

Proof. Observe that the columns of $\boldsymbol{A}$ constitute a basis of $\operatorname{null}\left(\mathbf{c}^{\mathrm{T}}\right)$, while the columns of $\boldsymbol{B}_{\boldsymbol{s}'}$ constitute a basis of $\operatorname{null}\left(\boldsymbol{d}^{\mathrm{T}}\right)$.

First, let $\mathbf{c}^{\mathrm{T}}\boldsymbol{d} \neq 0$. Then

$$\begin{aligned}
\operatorname{rank}\left(\boldsymbol{A}^{\mathrm{T}}\boldsymbol{B}_{\boldsymbol{s}'}\right) < K - 1 &\implies \exists \boldsymbol{v} \neq \boldsymbol{0} : \boldsymbol{A}^{\mathrm{T}}\boldsymbol{B}_{\boldsymbol{s}'}\boldsymbol{v} = \boldsymbol{0} \\
&\implies \boldsymbol{B}_{\boldsymbol{s}'}\boldsymbol{v} \in \operatorname{null}\left(\boldsymbol{A}^{\mathrm{T}}\right) \setminus \{\boldsymbol{0}\} \\
&\implies \exists \alpha \neq 0 : \boldsymbol{B}_{\boldsymbol{s}'}\boldsymbol{v} = \alpha\boldsymbol{c} \\
&\implies \mathbf{c}^{\mathrm{T}}\boldsymbol{d} = 0,
\end{aligned}$$

a contradiction.

Second, let $\operatorname{rank}\left(\boldsymbol{A}^{\mathrm{T}}\boldsymbol{B}_{\boldsymbol{s}'}\right) = K - 1$. Then

$$\begin{aligned}
\mathbf{c}^{\mathrm{T}}\boldsymbol{d} = 0 &\implies \exists \boldsymbol{v} \neq \boldsymbol{0} : \boldsymbol{c} = \boldsymbol{B}_{\boldsymbol{s}'}\boldsymbol{v}, \\
&\implies \boldsymbol{A}^{\mathrm{T}}\boldsymbol{B}_{\boldsymbol{s}'}\boldsymbol{v} = \boldsymbol{0} \\
&\implies \operatorname{rank}\left(\boldsymbol{A}^{\mathrm{T}}\boldsymbol{B}_{\boldsymbol{s}'}\right) < K - 1,
\end{aligned}$$

a contradiction. □

Lemma 3.5.8 provides a condition under which the product of two matrices has full rank. This condition is needed in the next proposition.

Proposition 3.5.9. *Let n denote the field of normal vectors of $\mathcal{M}$. Assume that*

$$n(s)^{\mathrm{T}} n(s') \neq 0, \forall s' \in \mathcal{M}. \tag{3.107}$$

Then $(\mathcal{U}, \psi_s)$ is a global parameterization of $\mathcal{M}$ and ψ_s is of class C^1.

Proof. By Lemma 3.5.7, ψ_s is a bijection. Hence, ψ_s is a global parameterization if both ψ_s and ψ_s^{-1} are continuous. Continuity of the inverse map ψ_s^{-1} follows directly from (3.106).

Continuity and differentiability of ψ_s depend on the properties of the function δ. Let $s' \in \mathcal{M}$. By assumption, $\mathcal{M}$ is a C^1 submanifold of $\mathbb{R}^K$. Consequently, there exists a local parameterization $(\mathcal{X}, \rho)$ of $\mathcal{M}$ such that $s' \in \rho(\mathcal{X})$ and ρ is of class C^1. Let

$$x = \rho^{-1}(s')$$

and

$$B_{s'} = \nabla \rho(x)^{\mathrm{T}}.$$

By Propositions A.8.3 and A.8.4,

$$\operatorname{rank}(B_{s'}) = K - 1,$$
$$n(s')^{\mathrm{T}} B_{s'} = \mathbf{0}.$$

By Lemma 3.5.7, there exists (μ, γ) such that

$$\rho(x) = s + Q_s \mu + n(s)\gamma. \tag{3.108}$$

Left-multiplying by Q_s^{T} and $n(s)^{\mathrm{T}}$ yields

$$\mu = Q_s^{\mathrm{T}}(\rho(x) - s), \tag{3.109}$$
$$\gamma = n(s)^{\mathrm{T}}(\rho(x) - s). \tag{3.110}$$

Equations (3.109) and (3.110) describe μ and γ as continuously differentiable functions of x:

$$\mu = f(x),$$
$$\gamma = g(x).$$

In order to have γ as a function of μ, the inverse of the function f is needed: Provided f^{-1} exists, we obtain $\delta(\mu) = g(f^{-1}(\mu))$. Observe that

$$\nabla \boldsymbol{f}(\boldsymbol{x})^{\mathrm{T}} = \boldsymbol{Q}_{\boldsymbol{s}}^{\mathrm{T}} \boldsymbol{B}_{\boldsymbol{s}'} \in \mathbb{R}^{K-1 \times K-1}.$$

Moreover, recall that the columns of $\boldsymbol{Q}_{\boldsymbol{s}}$ and $\boldsymbol{B}_{\boldsymbol{s}'}$ constitute a basis of the tangent space of $\mathcal{M}$ at $\boldsymbol{s}$ and $\boldsymbol{s}'$, respectively. Hence,

$$\boldsymbol{n}(\boldsymbol{s})^{\mathrm{T}} \boldsymbol{Q}_{\boldsymbol{s}} = \boldsymbol{0},$$
$$\boldsymbol{n}(\boldsymbol{s}')^{\mathrm{T}} \boldsymbol{B}_{\boldsymbol{s}'} = \boldsymbol{0}.$$

By Lemma 3.5.8, $\boldsymbol{n}(\boldsymbol{s})^{\mathrm{T}} \boldsymbol{n}(\boldsymbol{s}') \neq 0$ implies that $\nabla \boldsymbol{f}(\boldsymbol{x}) = \boldsymbol{Q}_{\boldsymbol{s}}^{\mathrm{T}} \boldsymbol{B}_{\boldsymbol{s}'}$ has full rank. Hence, according to the Inverse Function Theorem [66], $\boldsymbol{f}$ is invertible in some neighborhood of $\boldsymbol{x}$, and the inverse function $\boldsymbol{f}^{-1}$ is continuously differentiable, with

$$\nabla(\boldsymbol{f}^{-1})(\boldsymbol{\mu}) = (\nabla \boldsymbol{f}(\boldsymbol{x}))^{-1}.$$

Accordingly, in a neighborhood of $\boldsymbol{\mu}$,

$$\delta = g \circ \boldsymbol{f}^{-1},$$

where both g and $\boldsymbol{f}^{-1}$ are continuously differentiable functions, and

$$\begin{aligned} \nabla \delta(\boldsymbol{\mu}) &= (\nabla \boldsymbol{f}(\boldsymbol{x}))^{-1} \nabla g(\boldsymbol{x}) \\ &= (\boldsymbol{B}_{\boldsymbol{s}'}^{\mathrm{T}} \boldsymbol{Q}_{\boldsymbol{s}})^{-1} \boldsymbol{B}_{\boldsymbol{s}'}^{\mathrm{T}} \boldsymbol{n}(\boldsymbol{s}). \end{aligned} \tag{3.111}$$

Provided that

$$\boldsymbol{n}(\boldsymbol{s})^{\mathrm{T}} \boldsymbol{n}(\boldsymbol{s}') \neq 0, \forall \boldsymbol{s}' \in \mathcal{M},$$

above development holds for any $\boldsymbol{s}' \in \mathcal{M}$, implying that δ is continuously differentiable on $\mathcal{U}$. □

Proposition 3.5.9 shows that $(\mathcal{U}, \boldsymbol{\psi}_{\boldsymbol{s}})$ is a differentiable global parameterization of $\mathcal{M}$, provided that Condition (3.107) holds true. Noting that $\boldsymbol{n}(\boldsymbol{s}') \in \mathbb{R}_{+}^{K}, \forall \boldsymbol{s}' \in \mathcal{M}$, Condition (3.107) is fulfilled by choosing a point $\boldsymbol{s} = \boldsymbol{\psi}_{\boldsymbol{s}}(\boldsymbol{0})$ satisfying

$$\boldsymbol{n}(\boldsymbol{s}) \in \mathbb{R}_{++}^{K}.$$

In order to use the parameterization $(\mathcal{U}, \boldsymbol{\psi}_{\boldsymbol{s}})$ in a numerical algorithm for computing a KKT point, a method to compute $\boldsymbol{\psi}_{\boldsymbol{s}}(\boldsymbol{\mu})$ and $\nabla \boldsymbol{\psi}_{\boldsymbol{s}}(\boldsymbol{\mu})$ is needed, cf. Sect. 3.5.2. According to (3.104), evaluating $\boldsymbol{\psi}_{\boldsymbol{s}}(\boldsymbol{\mu})$ basically amounts to solving an intersection problem. In the following, it is shown that $\nabla \boldsymbol{\psi}_{\boldsymbol{s}}(\boldsymbol{\mu})$ can be obtained if an *arbitrary* basis of $\mathcal{T}_{\mathcal{M}}(\boldsymbol{\psi}_{\boldsymbol{s}}(\boldsymbol{\mu}))$ is known.

According to (3.105) and (3.111), the gradient matrix of $\boldsymbol{\psi}_{\boldsymbol{s}}$ at $\boldsymbol{\mu}$ is given by

$$\nabla\psi_s(\mu) = Q_s^T + \nabla\delta(\mu)n(s)^T \tag{3.112}$$

$$= Q_s^T + (B_{s'}^T Q_s)^{-1} B_{s'}^T n(s)n(s)^T$$

$$= Q_s^T + (B_{s'}^T Q_s)^{-1} B_{s'}^T - (B_{s'}^T Q_s)^{-1} B_{s'}^T Q_s Q_s^T$$

$$= (B_{s'}^T Q_s)^{-1} B_{s'}^T, \tag{3.113}$$

where the third equality follows from the identity

$$n(s)n(s)^T = I - Q_s Q_s^T.$$

The expression for $\nabla\psi(\mu)$ in (3.113) contains the matrix $B_{s'}$. However, complete knowledge of this matrix is not required: Let $C \in \mathbb{R}^{K \times K-1}$ denote a matrix such that

$$\text{range}\,(C) = \text{range}\,(B_{s'}) = \mathcal{T}_{\mathcal{M}}(s').$$

Then there exists a non-singular matrix $A \in \mathbb{R}^{K-1 \times K-1}$ such that $B_{s'} = CA^T$. With (3.113),

$$\nabla\psi_s(\mu) = (AC^T Q_s)^{-1} AC^T$$

$$= (C^T Q_s)^{-1} C^T. \tag{3.114}$$

It can be observed that $\nabla\psi_s(\mu)$ is independent of A. Hence, $\nabla\psi_s(\mu)$ depends only on Q_s and the tangent space at s', but not on the choice of the basis of $\mathcal{T}_{\mathcal{M}}(s')$. As an example, the gradient matrix at $\mu = \mathbf{0}$ can be obtained by choosing $C = Q_s$:

$$\nabla\psi_s(\mathbf{0}) = Q_s^T. \tag{3.115}$$

Equation (3.114) can also be obtained by a geometric argument. Assume μ is changed to $\mu + \Delta\mu$. In first order approximation around μ,

$$\psi(\mu + \Delta\mu) \approx s' + \Delta s,$$

with

$$\Delta s = \nabla\psi_s(\mu)^T \Delta\mu. \tag{3.116}$$

Note that $\Delta s \in \mathcal{T}_{\mathcal{M}}(s')$. The corresponding change in $\mathcal{T}_{\mathcal{M}}(s)$ is given by

$$\Delta v = Q_s \Delta\mu. \tag{3.117}$$

The tangent vectors Δs and Δv are illustrated in Fig. 3.5.

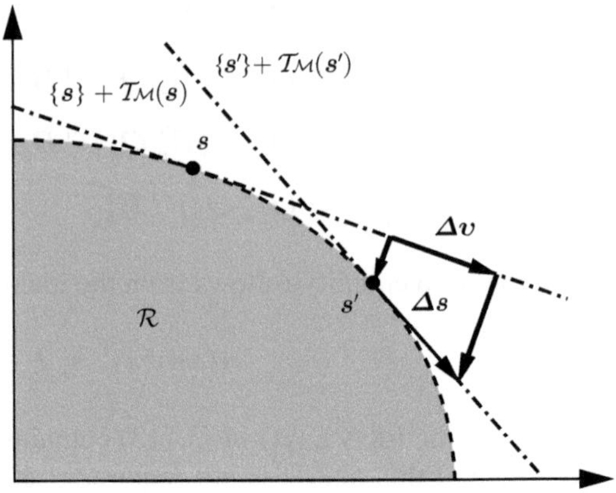

Fig. 3.5 The vector Δs is obtained by projecting Δv along range $(n(s))$ onto $\mathcal{T}_{\mathcal{M}}(s')$

Evidently, there exists a $P \in \mathbb{R}^{K\times K}$ that maps a change in $\mathcal{T}_{\mathcal{M}}(s)$ into a change in $\mathcal{T}_{\mathcal{M}}(s')$:

$$\Delta s = P\Delta v. \tag{3.118}$$

In fact, P is the non-orthogonal projection along range $(n(s))$ onto $\mathcal{T}_{\mathcal{M}}(s')$. Plugging (3.117) into (3.118) and comparing with (3.116) yields:

$$\nabla\psi(\mu)^{\mathrm{T}} = PQ_s.$$

A derivation of P is straightforward: By (3.116) and (3.112),

$$\begin{aligned}\Delta s &= Q_s\Delta\mu + n(s)\nabla\delta(\mu')^{\mathrm{T}}\Delta\mu\\ &= \Delta v + n(s)\nabla\delta(\mu')^{\mathrm{T}}\Delta\mu.\end{aligned}$$

Left-multiplying by $Q_sQ_s^{\mathrm{T}}$ yields

$$\Delta v = Q_sQ_s^{\mathrm{T}}\Delta s,$$

i.e., Δv is the orthogonal projection of Δs onto $\mathcal{T}_{\mathcal{M}}(s)$. In order to obtain Δs as a (non-orthogonal) projection of Δv, expand Δs in an arbitrary basis C of $\mathcal{T}_{\mathcal{M}}(s')$:

$$\Delta v = Q_sQ_s^{\mathrm{T}}C\Delta y.$$

Left-multiplying by Q_s^{T} and re-arranging yields

$$\Delta y = (Q_s^{\mathrm{T}}C)^{-1}Q_s^{\mathrm{T}}\Delta v,$$

implying

$$\begin{aligned}\Delta s &= C\Delta y \\ &= C(Q_s^{\mathrm{T}}C)^{-1}Q_s^{\mathrm{T}}\Delta v \\ &= P\Delta v.\end{aligned}$$

As a result,

$$P = C(Q_s^{\mathrm{T}}C)^{-1}Q_s^{\mathrm{T}},$$

which finally yields the expected result

$$\nabla\psi(\mu)^{\mathrm{T}} = C(Q_s^{\mathrm{T}}C)^{-1}.$$

To summarize, $(\mathcal{U}, \psi_s)$ is a differentiable global parameterization. In order to evaluate ψ_s at μ, a means to compute $\delta(\mu)$ is required. Computing $\delta(\mu)$ constitutes an intersection problem, cf. (3.104). In addition, computing the gradient $\nabla\psi_s(\mu)$ requires knowledge of a basis C of $\mathcal{T}_{\mathcal{M}}(\psi_s(\mu))$. If the normal vector $n(\psi_s(\mu))$ is available, such a basis can be obtained by computing a basis of null $\left(n(\psi_s(\mu))^{\mathrm{T}}\right)$. In comparison with the solution methods discussed in Sects. 3.3 and 3.4, the local method developed in this section is more demanding with respect to the elementary algorithms needed to operate in the rate space: The Lagrange duality-based approach requires an algorithm for maximizing a weighted sum of rates, while the polyblock algorithm needs a method for solving intersection problems. In contrast, the local method requires two elementary methods: one for solving intersection problems, and one for computing a tangent space basis.

3.5.4 A Varying Parameterization Approach

In Sect. 3.5.2, it is shown how a KKT point of the rate space problem (3.87) can be computed using standard algorithms, by employing a single global parameterization for the Pareto manifold $\mathcal{M}$. In Sect. 3.5.3, a generic global parameterization is proposed, together with a means to evaluate its gradient matrix. The parameterization is obtained by picking a point $s \in \mathcal{M}$ such that Condition (3.107) holds:

$$b_s(s') \neq 0, \forall s' \in \mathcal{M},$$

with a function $b_s : \mathcal{M} \to \mathbb{R}$,

$$b_s(s') = n(s)^{\mathrm{T}} n(s').$$

Using a single parameterization, it is straightforward to apply a standard algorithm. However, this approach has two drawbacks: first, it may not be obvious how to pick a point that satisfies Condition (3.107) (although finding a point with $\boldsymbol{n}(\boldsymbol{s}) \in \mathbb{R}_{++}^K$ is sufficient). Second, while $b_{\boldsymbol{s}}(\boldsymbol{s}') \neq 0$ is sufficient from a theoretical perspective, small values of $b_{\boldsymbol{s}}(\boldsymbol{s}')$ can lead to numerical problems. Consider the matrix $\boldsymbol{B}_{\boldsymbol{s}'}^{\mathrm{T}}\boldsymbol{Q}_{\boldsymbol{s}}$ in (3.111). According to Lemma 3.5.8, the matrix $\boldsymbol{B}_{\boldsymbol{s}'}^{\mathrm{T}}\boldsymbol{Q}_{\boldsymbol{s}}$ has full rank if $b_{\boldsymbol{s}}(\boldsymbol{s}') \neq 0$, while it is singular if $b_{\boldsymbol{s}}(\boldsymbol{s}') = 0$. Hence, the condition number of $\boldsymbol{B}_{\boldsymbol{s}'}^{\mathrm{T}}\boldsymbol{Q}_{\boldsymbol{s}}$ goes to infinity as $b(\boldsymbol{s}')$ approaches 0. As a result, for small values of $b(\boldsymbol{s}')$, small deviations in $\boldsymbol{\mu}$ lead to large deviations in $\delta(\boldsymbol{\mu})$—in other words, the problem to compute $\delta(\boldsymbol{\mu})$ (and thereby $\psi_{\boldsymbol{s}}(\boldsymbol{\mu})$) is ill-conditioned for small $b_{\boldsymbol{s}}(\boldsymbol{s}')$.

Both drawbacks can be circumvented by switching between parameterizations during the numerical maximization.

Proposition 3.5.10. *Let $\boldsymbol{s} \in \mathcal{M}$ and $0 < \beta < 1$. Then there exists $\alpha > 0$ such that*

$$b_{\boldsymbol{s}}(\boldsymbol{s}') > \beta, \forall \boldsymbol{s}' \in \mathcal{B}_\alpha(\boldsymbol{s}) \cap \mathcal{M}.$$

Proof. Note that $b_{\boldsymbol{s}}(\boldsymbol{s}) = \boldsymbol{n}(\boldsymbol{s})^{\mathrm{T}}\boldsymbol{n}(\boldsymbol{s}) = 1$. Moreover, $\mathcal{M}$ is a C^1 submanifold of $\mathbb{R}^K$. By Proposition A.8.4, the normal field $\boldsymbol{n}$ is a continuous function, implying that $b_{\boldsymbol{s}}$ is a continuous function. Thus, the preimage $b_{\boldsymbol{s}}^{-1}(\{\beta\} + \mathbb{R}_{++})$ is a nonempty open set (in the subspace topology on $\mathcal{M}$), with

$$\boldsymbol{s} \in b_{\boldsymbol{s}}^{-1}(\{\beta\} + \mathbb{R}_{++}). \tag{3.119}$$

As a result, there exists $\alpha > 0$ such that $\mathcal{B}_\alpha(\boldsymbol{s}) \cap \mathcal{M} \subset b_{\boldsymbol{s}}^{-1}(\{\beta\} + \mathbb{R}_{++})$. □

Proposition 3.5.10 states that at every point $\boldsymbol{s} \in \mathcal{M}$, there exists a neighborhood $\mathcal{N}_{\boldsymbol{s}}$ of $\boldsymbol{s}$ such that the condition number of the matrix $\boldsymbol{B}_{\boldsymbol{s}'}^{\mathrm{T}}\boldsymbol{Q}_{\boldsymbol{s}}$ is sufficiently small for all $\boldsymbol{s}' \in \mathcal{N}_{\boldsymbol{s}}$.

Now assume a standard numerical algorithm is used to find a KKT point of Problem (3.102). The algorithm generates a sequence of points $\{\boldsymbol{\mu}^{(n)}\}$ using a parameterization $\psi_{\boldsymbol{s}}$, starting at $\boldsymbol{\mu}^{(0)} = \boldsymbol{0}$. If at iteration n the value $b_{\boldsymbol{s}}(\psi_{\boldsymbol{s}}(\boldsymbol{\mu}^{(n)}))$ falls below a pre-defined value β, the parameterization is switched by setting $\boldsymbol{s}' = \psi_{\boldsymbol{s}}(\boldsymbol{\mu}^{(n)})$ and using the parameterization $\psi_{\boldsymbol{s}'}$ for the next iterations. The algorithm is restarted at $\boldsymbol{\mu}^{(0)} = \boldsymbol{0}$, but now using the parameterization $\psi_{\boldsymbol{s}'}$. Thus, in rate space, the algorithm continues at $\boldsymbol{s}'$. This process is repeated until the algorithm converges in one of the parameterizations.

It is important to note that switching between parameterizations leads to an interplay between choice of parameterization and the standard numerical algorithm that may impact the convergence properties of the overall algorithm. Assume that for m different parameterizations, a standard algorithm generates m sequences of iterates, and each sequence converges to a KKT point. Then switching between these parameterizations amounts to selecting subsequences from the m sequences, which does *not* imply that the overall sequence converges to a KKT point. Consider the following example:

$$x_n^m = \begin{cases} m^{-1}, & m = n, \\ 1, & m \neq n. \end{cases}$$

While every sequence $\{x_n^m\}_{n=1}^{\infty}$ converges to 1, the sequence $\{x_n^n\}_{n=1}^{\infty}$ converges to 0.

If the parameterization is switched a finite number of times, however, the convergence properties of the standard algorithm are preserved, simply due to the fact that after a certain number of iterations, the parameterization is kept fixed, as otherwise the number of switches would go to infinity with $n \to \infty$. Based on this argument, it is not necessarily a good choice to switch the parameterization at each iteration, despite the fact that this approach yields the largest value of $b_{\boldsymbol{s}}(\boldsymbol{s}')$. If the parameterization is switched at each iteration, additional considerations concerning the convergence of the overall algorithm are clearly necessary.

3.5.5 Boundary Points

In Sects. 3.5.1–3.5.4, a method to compute a KKT point of the rate space problem (3.87) is developed, under the assumption that

$$\mathcal{H} \subset \mathbb{R}_{++}^K.$$

Based on this assumption, it is sufficient to consider the set

$$\mathcal{M} = \operatorname{par}(\mathcal{R}) \cap \mathbb{R}_{++}^K.$$

In this section, the proposed method is extended such that it is possible to find KKT points of the rate space problem on the boundary of $\mathbb{R}_+^K$. Two cases can be distinguished: If a C^1 submanifold $\mathcal{M}$ *and* a global parameterization of $\mathcal{M}$ are available, with $\mathcal{M}$ satisfying

$$\mathcal{M} \cap \mathcal{H} = \operatorname{par}(\mathcal{R}) \cap \mathcal{H},$$

then the method from Sect. 3.5.2 can be applied without any further modifications. In general, though, a global parameterization that covers $\operatorname{par}(\mathcal{R}) \cap \mathcal{H}$ may not be available. In this case, an extension of the method developed in Sects. 3.5.1–3.5.4 is necessary.

Note that on the boundary of $\mathbb{R}_+^K$, the rate of at least one link is zero. In the following, links that have zero rate are denoted as *deactivated links*. Let $\mathcal{D}$ denote a set of link indices, i.e., $\mathcal{D} \subseteq \{1, \ldots, K\}$. For any such set $\mathcal{D}$, define a set $\mathcal{Z}_{\mathcal{D}}$ as follows:

$$\mathcal{Z}_{\mathcal{D}} = \left\{\boldsymbol{s} \in \mathbb{R}_+^K : s_k = 0, k \in \mathcal{D}\right\} \cap \left\{\boldsymbol{s} \in \mathbb{R}_+^K : s_k > 0, k \in \{1, \ldots, K\} \setminus \mathcal{D}\right\}.$$

A rate vector is element of $\mathcal{Z}_\mathcal{D}$ if any link k with $k \in \mathcal{D}$ has zero rate and all other links have a rate larger than zero. In other words, $\mathcal{D}$ contains the indices of the deactivated links. If all links are deactivated, $\mathcal{D} = \{1, \ldots, K\}$ and $\mathcal{Z}_{\{1,\ldots,K\}} = \{\mathbf{0}\}$. Likewise, if no link is deactivated, $\mathcal{D} = \emptyset$ and $\mathcal{Z}_\emptyset = \mathbb{R}^K_{++}$.

The extension of the local method for computing stationary points to points on the boundary of par $(\mathcal{R})$ is motivated by the following example:

Example 3.5.2. Let $K = 3$ and assume the rate region is the following slice of the unit ball:

$$\mathcal{R} = \left\{ \boldsymbol{s} \in \mathbb{R}^3_+ : \|\boldsymbol{s}\|_2 \leq 1 \right\}.$$

Clearly, $\mathcal{R}$ is proper. Moreover,

$$\text{par}\,(\mathcal{R}) = \left\{ \boldsymbol{s} \in \mathbb{R}^3_+ : \|\boldsymbol{s}\|_2 = 1 \right\}.$$

The Pareto set par $(\mathcal{R})$ can be decomposed into a set of disjoint submanifolds of $\mathbb{R}^3$:

$$\text{par}\,(\mathcal{R}) = \mathcal{M}_\emptyset \cup \mathcal{M}_{\{1\}} \cup \mathcal{M}_{\{2\}} \cup \mathcal{M}_{\{3\}} \cup \mathcal{M}_{\{1,2\}} \cup \mathcal{M}_{\{1,3\}} \cup \mathcal{M}_{\{2,3\}},$$

where

$$\mathcal{M}_\mathcal{D} = \text{par}\,(\mathcal{R}) \cap \mathcal{Z}_\mathcal{D}.$$

This decomposition is illustrated in Fig. 3.6. For example, $\mathcal{M}_\emptyset = \text{par}\,(\mathcal{R}) \cap \mathbb{R}^3_{++}$,

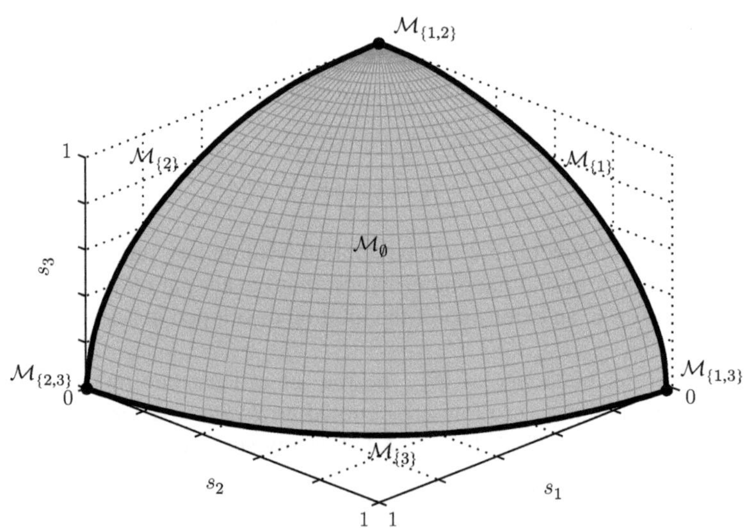

Fig. 3.6 A decomposition of the Pareto set into a set of disjoint manifolds

$$\mathcal{M}_{\{1\}} = \left\{ s \in \mathbb{R}_+^3 : s_1 = 0, s_2 > 0, s_3 > 0, \|s\|_2 = 1 \right\},$$

and

$$\begin{aligned}\mathcal{M}_{\{2,3\}} &= \left\{ s \in \mathbb{R}_+^3 : s_1 > 0, s_2 = 0, s_3 = 0, \|s\|_2 = 1 \right\} \\ &= \{(1,0,0)\}.\end{aligned}$$

The manifold $\mathcal{M}_\emptyset$ is a 2-dimensional C^1 submanifold of $\mathbb{R}^3$, while $\mathcal{M}_{\{1\}}$, $\mathcal{M}_{\{2\}}$, and $\mathcal{M}_{\{3\}}$ are C^1 submanifolds of dimension 1. The 0-dimensional submanifolds $\mathcal{M}_{\{1,2\}}$, $\mathcal{M}_{\{1,3\}}$, and $\mathcal{M}_{\{2,3\}}$ correspond to the single-link points of $\mathcal{R}$.

Example 3.5.2 directly motivates the following assumption:

Assumption 3.5.6. *The Pareto set of the rate region $\mathcal{R}$ can be decomposed as follows:*

$$\mathrm{par}\,(\mathcal{R}) = \bigcup_{\mathcal{D} \subset \{1,\ldots,K\}} \mathcal{M}_\mathcal{D},$$

where

$$\mathcal{M}_\mathcal{D} = \mathrm{par}\,(\mathcal{R}) \cap \mathcal{Z}_\mathcal{D} \tag{3.120}$$

is a C^1 submanifold of $\mathbb{R}^K$ *of dimension $K - 1 - \mathrm{card}\,(\mathcal{D})$.*

By assumption, each manifold $\mathcal{M}_\mathcal{D}$ offers the differentiable structure needed to compute a stationary point in $\mathcal{M}_\mathcal{D}$ using a numerical algorithm. This structure, however, is limited to each manifold. A single global differentiable structure that covers par $(\mathcal{R})$ is not available. As a consequence, a method is needed to connect the separate structures such that an algorithm can move from one manifold to another in a well-defined manner.

Note that the set $\mathcal{D}$ contains the indices of the links that have zero rate for all points in $\mathcal{M}_\mathcal{D}$. Accordingly, in $\mathcal{M}_\emptyset$, all links are active. Moving between manifolds is based on the following idea: Start the search for a stationary point in $\mathcal{M}_\emptyset$ (all links active). If there is an indication that a better performance may be achieved by deactivating some links, add the indices of these links to $\mathcal{D}$ and continue the search for a stationary point in $\mathcal{M}_\mathcal{D}$. If necessary, deactivate further links, thereby moving to the next manifold, until a stationary point is found and no further deactivation of users seems necessary. In order to turn this basic idea into a complete method, a number of problems need to be addressed:

(1) How can it be ensured that searching for a stationary point in $\mathcal{M}_\mathcal{D} \cap \mathcal{H}$ is a well-posed problem?
(2) How to compute a stationary point in $\mathcal{M}_\mathcal{D} \cap \mathcal{H}$? Sections 3.5.1–3.5.4 provide a method to compute a stationary point in $\mathcal{M}_\emptyset \cap \mathcal{H}$, but this method uses the field

of normal vectors $\boldsymbol{n}$, which is only defined for a $K-1$-dimensional submanifold of $\mathbb{R}^K$.

(3) What to use as an indicator that a link should be deactivated?

Solutions to all three problems are discussed next.

3.5.5.1 Ensuring Existence of a Stationary Point

It is a nontrivial problem to guarantee that searching for a stationary point in $\mathcal{M}_{\mathcal{D}} \cap \mathcal{H}$ is a well-posed problem. Well-posedness of searching for a stationary point is closely related to the problem whether there exists $\boldsymbol{s}^* \in \mathcal{M}_{\mathcal{D}} \cap \mathcal{H}$ such that

$$u(\boldsymbol{s}^*) = \sup\{u(\boldsymbol{s}) : \boldsymbol{s} \in \mathcal{M}_{\mathcal{D}} \cap \mathcal{H}\}.$$

The manifolds $\mathcal{M}_{\mathcal{D}}$ are not closed, thus it is not clear whether $\mathcal{M}_{\mathcal{D}} \cap \mathcal{H}$ is closed. As a result, it cannot be assumed that the supremum is attained in $\mathcal{M}_{\mathcal{D}} \cap \mathcal{H}$. Due to the fact that the supremum is not necessarily attained in $\mathcal{M}_{\mathcal{D}} \cap \mathcal{H}$, it may not be possible to find a stationary point in $\mathcal{M}_{\mathcal{D}} \cap \mathcal{H}$. As an example, assume an ascent method is used, starting from an initial value $\boldsymbol{s}^{(0)}$, with the intention to compute a stationary point in $\mathcal{M}_{\mathcal{D}}$. This approach may fail, as the level set $u^{-1}(\{u(\boldsymbol{s}^{(0)})\} + \mathbb{R}_+)$ may simply not contain any stationary points. In contrast, if the maximum is attained in $\mathcal{M}_{\mathcal{D}}$, the level set $u^{-1}(\{u(\boldsymbol{s}^{(0)})\} + \mathbb{R}_+)$ is guaranteed to contain at least one stationary point.

In Sect. 3.5.1, it is assumed that $\mathcal{H} \subset \mathbb{R}^K_{++}$. Based on this assumption,

$$\mathcal{M}_{\emptyset} \cap \mathcal{H} = \mathrm{par}\,(\mathcal{R}) \cap \mathcal{H}. \tag{3.121}$$

By Proposition A.3.4 and Assumption 3.5.3, the set par $(\mathcal{R})$ is compact. Moreover, $\mathcal{H}$ is closed, thus (3.121) implies that $\mathcal{M}_{\emptyset} \cap \mathcal{H}$ is compact, and the maximum is attained in $\mathcal{M}_{\emptyset} \cap \mathcal{H}$. As shown in Sect. 3.5.1, the assumption $\mathcal{H} \subset \mathbb{R}^K_{++}$ can be fulfilled by replacing $\mathcal{H}$ by $\mathcal{H} \cap \mathcal{A}$, where

$$\mathcal{A} = \{\eta \mathbf{1}\} + \mathbb{R}^K_+,$$

and $\eta > 0$. A similar approach can be applied to $\mathcal{M}_{\mathcal{D}} \cap \mathcal{H}$. Define

$$\mathcal{A}_{\mathcal{D}} = \left\{\boldsymbol{s} \in \mathbb{R}^K_+ : s_k \geq \eta, k \in \{1, \ldots, K\} \setminus \mathcal{D}\right\}.$$

The set $\mathcal{A}_{\mathcal{D}}$ requires that all active links (i.e., all links not in $\mathcal{D}$) have a rate of at least η. By (3.120),

$$\begin{aligned}\mathcal{M}_{\mathcal{D}} \cap \mathcal{A}_{\mathcal{D}} &= \mathrm{par}\,(\mathcal{R}) \cap \mathcal{Z}_{\mathcal{D}} \cap \mathcal{A}_{\mathcal{D}} \\ &= \mathrm{par}\,(\mathcal{R}) \cap \left\{\boldsymbol{s} \in \mathbb{R}^K_+ : s_k = 0, k \in \mathcal{D}\right\} \cap \\ &\quad \left\{\boldsymbol{s} \in \mathbb{R}^K_+ : s_k \geq \eta, k \in \{1, \ldots, K\} \setminus \mathcal{D}\right\}.\end{aligned}$$

Again, according to Proposition A.3.4 and Assumption 3.5.3, the set $\operatorname{par}(\mathcal{R})$ is compact. Thus, $\mathcal{M}_{\mathcal{D}} \cap \mathcal{A}_{\mathcal{D}}$ is the intersection of a compact set with two closed sets, which is compact. Finally, the set $\mathcal{H}$ is closed, implying that $\mathcal{M}_{\mathcal{D}} \cap \mathcal{A}_{\mathcal{D}} \cap \mathcal{H}$ is compact, which implies the existence of a stationary point in $\mathcal{M}_{\mathcal{D}} \cap \mathcal{A}_{\mathcal{D}} \cap \mathcal{H}$.

3.5.5.2 An Indicator for Deactivating Links

Clearly, requiring a minimum rate may add stationary points on the boundary of $\mathcal{A}_{\mathcal{D}}$. Given $\boldsymbol{s} \in \mathcal{R}$, define a set $\mathcal{W}(\boldsymbol{s})$ as follows:

$$\mathcal{W}(\boldsymbol{s}) = \{k \in \{1, \dots, K\} : s_k = \eta\}.$$

Note that

$$\mathcal{W}(\boldsymbol{s}) \subseteq \{1, \dots, K\} \setminus \mathcal{D}, \forall \boldsymbol{s} \in \mathcal{Z}_{\mathcal{D}}.$$

In other words, for $\boldsymbol{s} \in \mathcal{Z}_{\mathcal{D}}$, $\mathcal{W}(\boldsymbol{s})$ contains the indices of the minimum rate constraints that are active at $\boldsymbol{s}$. Consider the case $\mathcal{D} = \emptyset$. Let $\boldsymbol{s}^*$ denote a stationary point in $\mathcal{M}_{\emptyset} \cap \mathcal{A}_{\emptyset} \cap \mathcal{H}$. Assume that the vectors in the set

$$\{\boldsymbol{n}(\boldsymbol{s}^*)\} \cup \{\nabla h_m(\boldsymbol{s}^*) : m \in \mathcal{J}(\boldsymbol{s}^*)\} \cup \{\boldsymbol{e}_k : k \in \mathcal{W}(\boldsymbol{s}^*)\}$$

are linearly independent. As discussed in Sect. 3.5.1, under the linear independence assumption, the tangent cone of $\mathcal{M}_{\emptyset} \cap \mathcal{A}_{\emptyset} \cap \mathcal{H}$ at $\boldsymbol{s}^*$ is given by

$$\mathcal{T}_{\mathcal{M}_{\emptyset} \cap \mathcal{A}_{\emptyset} \cap \mathcal{H}}(\boldsymbol{s}^*) = \operatorname{null}\left(\boldsymbol{n}(\boldsymbol{s}^*)^{\mathrm{T}}\right) \cap \bigcap_{m \in \mathcal{J}(\boldsymbol{s}^*)} \{\boldsymbol{v} : \nabla h_m(\boldsymbol{s}^*)^{\mathrm{T}} \boldsymbol{v} \geq 0\} \cap \bigcap_{k \in \mathcal{W}(\boldsymbol{s}^*)} \{\boldsymbol{v} : \boldsymbol{e}_k^{\mathrm{T}} \boldsymbol{v} \geq 0\}.$$

Moreover, every stationary point is a KKT point, thus there exists $\boldsymbol{\lambda} \in \mathbb{R}_+^M$, $\boldsymbol{\beta} \in \mathbb{R}_+^K$, and $\nu \in \mathbb{R}$ such that

$$\nabla u(\boldsymbol{s}^*) + \sum_{m \in \mathcal{J}(\boldsymbol{s}^*)} \lambda_m \nabla h_m(\boldsymbol{s}^*) + \sum_{k \in \mathcal{W}(\boldsymbol{s}^*)} \beta_k \boldsymbol{e}_k + \nu \boldsymbol{n}(\boldsymbol{s}^*) = \boldsymbol{0}. \tag{3.122}$$

Now assume that $\boldsymbol{s}^*$ is obtained by a numerical algorithm. If $\boldsymbol{s}^*$ is not a stationary point in $\mathcal{M}_{\emptyset} \cap \mathcal{H}$, then the algorithm converged to a stationary point that was created by adding the minimum rate constraint. Under the linear independence assumption, $\boldsymbol{s}^*$ is a stationary point in $\mathcal{M}_{\emptyset} \cap \mathcal{H}$ if and only if

$$\beta_k = 0, \forall k \in \mathcal{W}(\boldsymbol{s}^*).$$

If there exists $k \in \mathcal{W}(\boldsymbol{s}^*)$ such that $\beta_k > 0$, then it is possible to increase the utility value by removing the minimum rate constraint associated with link k—due to the fact that $\boldsymbol{s}^*$ is not a stationary point in $\mathcal{M}_\emptyset \cap \mathcal{H}$, there exists an ascent direction at $\boldsymbol{s}^*$. If η is sufficiently small, this result provides an indicator that it may be beneficial to deactivate all links $k \in \mathcal{W}(\boldsymbol{s}^*)$ with $\beta_k > 0$. Accordingly, the set $\mathcal{D}$ is chosen as follows:

$$\mathcal{D} = \{k \in \mathcal{W}(\boldsymbol{s}^*) : \beta_k > 0\}.$$

If $\mathcal{D}$ is nonempty, the algorithm continues by computing a stationary point in $\mathcal{M}_\mathcal{D} \cap \mathcal{A}_\mathcal{D} \cap \mathcal{H}$.

3.5.5.3 Computing a Stationary Point in the Submanifolds

In Sects. 3.5.1–3.5.4, a method for computing a stationary point in $\mathcal{M} \cap \mathcal{A} \cap \mathcal{H}$ is developed. Moreover, an indicator for moving from $\mathcal{M}$ to $\mathcal{M}_\mathcal{D}$ is available. The method from Sects. 3.5.1–3.5.4 relies on two properties:

(1) $\mathcal{M}$ is the Pareto manifold of a proper subset of $\mathbb{R}_+^K$.
(2) $\mathcal{M}$ is a C^1 submanifold of $\mathbb{R}^K$.

The combination of both properties implies that $\mathcal{M}$ is a $K-1$-dimensional submanifold of $\mathbb{R}^K$ for which a differentiable global parameterization exists. In contrast, a manifold $\mathcal{M}_\mathcal{D}$ has dimension $K-1-\mathrm{card}(\mathcal{D})$. In this section, it is shown that by describing a manifold $\mathcal{M}_\mathcal{D}$ as a submanifold of $\mathbb{R}^{K-\mathrm{card}(\mathcal{D})}$, the method from Sects. 3.5.1–3.5.4 can be used to compute a stationary point in $\mathcal{M}_\mathcal{D} \cap \mathcal{A}_\mathcal{D} \cap \mathcal{H}$, for arbitrary $\mathcal{D}$.

Recall that $\mathcal{M}_\mathcal{D} \subset \mathcal{Z}_\mathcal{D}$, i.e., when operating in $\mathcal{M}_\mathcal{D}$, the rates of $\mathrm{card}(\mathcal{D})$ links are always zero. Accordingly, these links can simply be removed from the description. In order to do so, let $\ell : \{1, \ldots, K - \mathrm{card}(\mathcal{D})\} \to \{1, \ldots, K\} \setminus \mathcal{D}$ be a bijection and define a matrix $\boldsymbol{F} \in \mathbb{R}^{K \times K-\mathrm{card}(\mathcal{D})}$ as follows:

$$\boldsymbol{F} = \left[\boldsymbol{e}_{\ell(1)}, \ldots, \boldsymbol{e}_{\ell(K-\mathrm{card}(\mathcal{D}))}\right].$$

The columns of $\boldsymbol{F}$ constitute an orthonormal basis of the linear subspace

$$\mathcal{E}_\mathcal{D} = \left\{\boldsymbol{s} \in \mathbb{R}^K : s_k = 0, k \in \mathcal{D}\right\}.$$

Define a linear map $\boldsymbol{\kappa}_\mathcal{D} : \mathcal{E}_\mathcal{D} \to \mathbb{R}^{K-\mathrm{card}(\mathcal{D})}$,

$$\boldsymbol{\kappa}_\mathcal{D}(\boldsymbol{s}) = \boldsymbol{F}^\mathrm{T}\boldsymbol{s}.$$

Note that $\boldsymbol{\kappa}_\mathcal{D}$ simply selects the nonzero entries from a rate vector $\boldsymbol{s} \in \mathcal{Z}_\mathcal{D}$ and discards all coordinates s_k with $k \in \mathcal{D}$. The effect of applying $\boldsymbol{\kappa}_\mathcal{D}$ to $\mathcal{M}_\mathcal{D}$ is illustrated in Fig. 3.7.

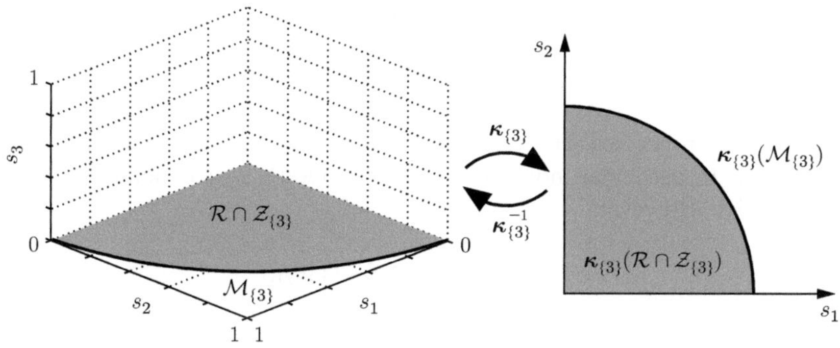

Fig. 3.7 Coordinates that are always zero are discarded by applying the map $\kappa_{\mathcal{D}}$

Moreover, $\kappa_{\mathcal{D}}$ is a C^{∞}-diffeomorphism, with

$$\kappa_{\mathcal{D}}^{-1}(y) = Fy.$$

Accordingly, $\kappa_{\mathcal{D}}(\mathcal{M}_{\mathcal{D}})$ is a $K - 1 - \text{card}(\mathcal{D})$-dimensional C^1 submanifold of $\mathbb{R}^{K-\text{card}(\mathcal{D})}$. Hence, showing the C^1 submanifold property is straightforward. In the next step, it is shown that $\kappa_{\mathcal{D}}(\mathcal{M}_{\mathcal{D}})$ is the Pareto manifold of a proper subset of $\mathbb{R}_+^{K-\text{card}(\mathcal{D})}$.

Figure 3.7 indicates that $\kappa_{\mathcal{D}}(\mathcal{M}_{\mathcal{D}})$ is the Pareto set of $\kappa_{\mathcal{D}}(\mathcal{R} \cap \mathcal{Z}_{\mathcal{D}})$. This is indeed true, as shown in the following: By definition of a Pareto set,

$$\text{par}\,(\mathcal{R}) \cap \mathcal{Z}_{\mathcal{D}} \subseteq \text{par}\,(\mathcal{R} \cap \mathcal{Z}_{\mathcal{D}})\,.$$

Moreover, by Proposition A.4.6,

$$\text{wpar}\,(\mathcal{R}) \cap \mathcal{Z}_{\mathcal{D}} \supseteq \text{par}\,(\mathcal{R} \cap \mathcal{Z}_{\mathcal{D}})\,.$$

By Assumption 3.5.3, $\text{wpar}\,(\mathcal{R}) = \text{par}\,(\mathcal{R})$, hence

$$\text{par}\,(\mathcal{R}) \cap \mathcal{Z}_{\mathcal{D}} = \text{par}\,(\mathcal{R} \cap \mathcal{Z}_{\mathcal{D}})\,.$$

Finally, note that for all $s, s' \in \mathcal{E}_{\mathcal{D}}$,

$$\kappa_{\mathcal{D}}(s) > \kappa_{\mathcal{D}}(s') \iff s > s',$$

which implies that $\kappa_{\mathcal{D}}(\text{par}\,(\mathcal{A})) = \text{par}\,(\kappa_{\mathcal{D}}(\mathcal{A}))$ for any $\mathcal{A} \subseteq \mathcal{E}_{\mathcal{D}}$. Thus,

$$\kappa_{\mathcal{D}}(\mathcal{M}_{\mathcal{D}}) = \kappa_{\mathcal{D}}(\text{par}\,(\mathcal{R} \cap \mathcal{Z}_{\mathcal{D}})) \tag{3.123}$$

$$= \text{par}\,(\kappa_{\mathcal{D}}(\mathcal{R} \cap \mathcal{Z}_{\mathcal{D}}))\,, \tag{3.124}$$

which is the expected result.

Recall the definition of $\mathcal{M}$ from Sect. 3.5.1:

$$\mathcal{M} = \text{par}\,(\mathcal{R}) \cap \mathbb{R}^K_{++},$$

where $\mathcal{R}$ is a proper set. The representation of $\boldsymbol{\kappa}_{\mathcal{D}}(\mathcal{M}_{\mathcal{D}})$ in (3.124) is similar, but different. In particular, note that the set $\boldsymbol{\kappa}_{\mathcal{D}}(\mathcal{R} \cap \mathcal{Z}_{\mathcal{D}})$ is neither closed nor comprehensive. However,

$$\mathcal{R} \cap \mathcal{Z}_{\mathcal{D}} = \mathcal{R} \cap \mathcal{E}_{\mathcal{D}} \cap \mathcal{Z}_{\mathcal{D}},$$

and

$$\text{par}\,(\mathcal{R} \cap \mathcal{E}_{\mathcal{D}} \cap \mathcal{Z}_{\mathcal{D}}) = \text{par}\,(\mathcal{R} \cap \mathcal{E}_{\mathcal{D}}) \cap \mathcal{Z}_{\mathcal{D}},$$

hence

$$\boldsymbol{\kappa}_{\mathcal{D}}(\mathcal{M}_{\mathcal{D}}) = \text{par}\,(\boldsymbol{\kappa}_{\mathcal{D}}(\mathcal{R} \cap \mathcal{E}_{\mathcal{D}})) \cap \mathbb{R}^{K-\text{card}(\mathcal{D})}_{++}. \tag{3.125}$$

The set $\boldsymbol{\kappa}_{\mathcal{D}}(\mathcal{R} \cap \mathcal{E}_{\mathcal{D}})$ is a proper subset of $\mathbb{R}^{K-\text{card}(\mathcal{D})}_{+}$. This finally shows that $\boldsymbol{\kappa}_{\mathcal{D}}(\mathcal{M}_{\mathcal{D}})$ has the same properties as $\mathcal{M}$, only the dimension is reduced from K to $K - \text{card}\,(\mathcal{D})$. As a result, a stationary point of the problem

$$\max_{\boldsymbol{y}} u(\boldsymbol{\kappa}_{\mathcal{D}}^{-1}(\boldsymbol{y})) \quad \text{s.t.} \quad \boldsymbol{y} \in \boldsymbol{\kappa}_{\mathcal{D}}(\mathcal{M}_{\mathcal{D}} \cap \mathcal{A}_{\mathcal{D}} \cap \mathcal{H})$$

can be computed using the framework developed in Sects. 3.5.1–3.5.4, and the indicator for deactivating links can be applied to such a stationary point in a straightforward manner. The final step is obvious from the construction of $\boldsymbol{\kappa}_{\mathcal{D}}$:

Proposition 3.5.11. *Let $\boldsymbol{y} \in \boldsymbol{\kappa}_{\mathcal{D}}(\mathcal{M}_{\mathcal{D}} \cap \mathcal{A}_{\mathcal{D}} \cap \mathcal{H})$. Then $\boldsymbol{y}$ is a stationary point in $\boldsymbol{\kappa}_{\mathcal{D}}(\mathcal{M}_{\mathcal{D}} \cap \mathcal{A}_{\mathcal{D}} \cap \mathcal{H})$ if and only if $\boldsymbol{\kappa}_{\mathcal{D}}^{-1}(\boldsymbol{y})$ is a stationary point in $\mathcal{M}_{\mathcal{D}} \cap \mathcal{A}_{\mathcal{D}} \cap \mathcal{H}$.*

Proof. Note that $\boldsymbol{\kappa}_{\mathcal{D}}^{-1}(\boldsymbol{y}) = \boldsymbol{F}\boldsymbol{y}$ and

$$\nabla(u \circ \boldsymbol{\kappa}_{\mathcal{D}}^{-1})(\boldsymbol{y}) = \boldsymbol{F}^{\mathrm{T}} \nabla u(\boldsymbol{F}\boldsymbol{y}).$$

Let $\boldsymbol{s} = \boldsymbol{F}\boldsymbol{y}$. Then

$$\begin{aligned}
& (\boldsymbol{F}^{\mathrm{T}} \nabla u(\boldsymbol{s}))^{\mathrm{T}} \boldsymbol{x} \leq \boldsymbol{0}, \forall \boldsymbol{x} \in \mathcal{T}_{\boldsymbol{\kappa}_{\mathcal{D}}(\mathcal{M}_{\mathcal{D}} \cap \mathcal{A}_{\mathcal{D}} \cap \mathcal{H})}(\boldsymbol{y}) \\
\Longleftrightarrow \quad & \nabla u(\boldsymbol{s})^{\mathrm{T}} \boldsymbol{F} \boldsymbol{x} \leq \boldsymbol{0}, \forall \boldsymbol{x} \in \boldsymbol{\kappa}_{\mathcal{D}}\big(\mathcal{T}_{\mathcal{M}_{\mathcal{D}} \cap \mathcal{A}_{\mathcal{D}} \cap \mathcal{H}}(\boldsymbol{s})\big) \\
\Longleftrightarrow \quad & \nabla u(\boldsymbol{s})^{\mathrm{T}} \boldsymbol{F} \boldsymbol{F}^{\mathrm{T}} \boldsymbol{v} \leq \boldsymbol{0}, \forall \boldsymbol{v} \in \mathcal{T}_{\mathcal{M}_{\mathcal{D}} \cap \mathcal{A}_{\mathcal{D}} \cap \mathcal{H}}(\boldsymbol{s}) \\
\Longleftrightarrow \quad & \nabla u(\boldsymbol{s})^{\mathrm{T}} \boldsymbol{v} \leq \boldsymbol{0}, \forall \boldsymbol{v} \in \mathcal{T}_{\mathcal{M}_{\mathcal{D}} \cap \mathcal{A}_{\mathcal{D}} \cap \mathcal{H}}(\boldsymbol{s}),
\end{aligned}$$

where the last equivalence follows from the fact that $\boldsymbol{F}\boldsymbol{F}^{\mathrm{T}}\boldsymbol{v} = \boldsymbol{\kappa}_{\mathcal{D}}^{-1}(\boldsymbol{\kappa}_{\mathcal{D}}(\boldsymbol{v})) = \boldsymbol{v}, \forall \boldsymbol{v} \in \mathcal{Z}_{\mathcal{D}}$. □

3.5.6 Related Work

Optimization problems of the form

$$\max_x f(x) \quad \text{s.t.} \quad x \in \mathcal{M}, \tag{3.126}$$

where f is real-valued and M is a differentiable manifold, constitute a special class of constrained optimization problems, known as *optimization on manifolds* (OM). OM is an active area of research that has attracted significant attention in the last two decades [67]. Two of the early works in this field are [68] and [59], which both use a Riemannian structure and the concept of geodesics to extend well-known methods from unconstrained optimization to the optimization of manifolds. In [68], differential geometric concepts are used to prove the convergence properties of Rosen's projected gradient algorithm in an idealized setting. Results in [68] are based on the assumption that the set of active constraints remains constant, which turns the inequality constrained problem into an equality constrained problem. Under a regularity assumption, the solution set of the equality constraints can then be described as a manifold. While the analysis in [68] is elegant, a main problem in proving the global convergence of Rosen's algorithm is to show that the set of active constraints actually remains constant, cf. [64]. Moreover, [68] performs a gradient descent along geodesics, but computing geodesic curves may be a hard problem for an arbitrary differentiable manifold $\mathcal{M}$. The author in [59] uses similar concepts to develop Newton methods for optimization on manifolds. Again, geodesics are used as the main tool in the theoretic analysis. In addition, [59] provides a practical method which basically corresponds to the parameterization used in Sect. 3.5.3. While computing geodesics may be straightforward for some manifolds [67], in other cases using the local parameterizations of the manifold may be a better choice [69]. If parameterizations are used, it is generally necessary to switch between parameterizations, due to the fact that a local parameterization does not cover the whole manifold in general. For Newton methods, such a *varying parameterization approach* is proposed in [70]. In [71], a gradient descent method on manifolds based on varying parameterizations is developed, and its global convergence as well as rate of convergence are shown.

Comparing the canonical OM problem (3.126) to the rate space problem (3.95) shows two significant differences: First, in Problem (3.95) there is an additional inequality constraint. Second, for the Pareto manifold $\mathcal{M}$ in Problem (3.95), a global parameterization that completely covers $\mathcal{M}$ is available. The presence of an additional inequality constraint is critical: almost all works on OM assume that there are no additional constraints (referred to as an "unconstrained optimization in a constrained search space" [67]). From this point of view, Problem (3.95) is harder than the usual optimization on manifolds problem, and it is not possible to simply apply algorithms from the OM literature, except for the special case that the inequality constraint can be ignored (which is true if it is known beforehand that the employed algorithm will only generate iterates in $\text{int}\,(\mathcal{H})$). Fortunately, the availability of a global parameterization actually makes Problem (3.95) easier to solve than the general OM problem. As shown in Sect. 3.5.2, by using the global parameterization, the rate space problem can be transformed into a standard inequality constrained problem, which is solvable by standard algorithms.

It is an interesting question under which conditions an OM problem (possibly with additional constraints) can be solved by the method developed in Sects. 3.5.1–3.5.4. Obviously, the key to using standard methods is the availability of a differentiable global parameterization. Section 3.5.3 provides a method to construct a global parameterization if $\mathcal{M}$ is a connected subset of the weak Pareto set of a proper rate region. By similar arguments, it can be shown that a connected manifold $\mathcal{M}$ has a global parameterization if there exists a compact set $\mathcal{R}$ such that $\mathcal{M} \subseteq \text{wpar}(\mathcal{R})$. Such a set exists if and only if $\mathcal{M} = \text{wpar}(\mathcal{M})$ and $\mathcal{M} \subset \mathbb{R}^K$ is bounded. As a result, standard methods for inequality constrained optimization can be applied if $\mathcal{M}$ is a differentiable submanifold of $\mathbb{R}^K$, $\mathcal{M}$ is bounded and connected, and $\mathcal{M} = \text{wpar}(\mathcal{M})$ (under the assumption that the OM problem is well-defined and the necessary constraint qualification holds true, cf. Sect. 3.5.1).

Chapter 4
Physical Layer Models

In the general model introduced in Chap. 2, the physical layer of a wireless communication system is described by a set of feasible physical layer parameter setups $\mathcal{X}_1$ and a rate function $\boldsymbol{r}$ that assigns a rate vector $\boldsymbol{r}(\boldsymbol{x}_1)$ to every feasible parameter setup $\boldsymbol{x}_1 \in \mathcal{X}_1$. In this chapter, two examples for such a pair $(\boldsymbol{r}, \mathcal{X}_1)$ are provided. In both examples, the network topology is simple: broadcast channels are considered, i.e., one centralized transmitter transmits separate information to multiple receivers over a single hop. The transmitter is equipped with multiple transmit antennas, allowing for spatial signal processing at the transmitter. Transmissions over different links are not separated by an orthogonal multiple access scheme. Instead, the optimal transmission scheme is determined by solving a utility maximization problem. Interference between links leads to nonconvex parameter space problems. While Sect. 4.1 assumes single antenna receivers, Sect. 4.2 discusses the case of multiple antennas at both the receivers and the transmitter. Moreover, while in Sect. 4.1, interference is only mitigated by linear precoding at the transmitter, in Sect. 4.2 the transmitter employs a precoding technique known as *dirty paper coding* (DPC) [72] to cancel known interference.

4.1 MISO Broadcast Channel with Linear Precoding

A multi-antenna broadcast channel is considered. One transmitter with N transmit antennas transmits to K single-antenna receivers. The received signal at the k-th receiver is given by

$$y_k = \sum_{q=1}^{K} \boldsymbol{h}_k^{\mathrm{H}} \boldsymbol{w}_q s_q + \eta_k,$$

where $\boldsymbol{h}_k^{\mathrm{H}} \in \mathbb{C}^{1\times N}$, $\boldsymbol{w}_k \in \mathbb{C}^{N\times 1}$, and $s_k \in \mathbb{C}$ are the channel, precoder, and data symbol of user k, respectively, and η_k is circularly symmetric AWGN with

J. Brehmer, *Utility Maximization in Nonconvex Wireless Systems*, Foundations in Signal Processing, Communications and Networking 5, DOI 10.1007/978-3-642-17438-4_4,

zero mean and variance σ^2. Throughout this work, it is assumed that $\sigma^2 > 0$. The transmitted signal

$$\boldsymbol{x} = \sum_{k=1}^{K} \boldsymbol{w}_k s_k$$

is subject to a sum-power constraint

$$\mathrm{E}\left[\|\boldsymbol{x}\|_2^2\right] \leq P. \tag{4.1}$$

4.1.1 PHY Layer Parameterization

Under the assumption of uncorrelated, unit power data symbols, the sum-power constraint (4.1) translates into a constraint on the precoders $\boldsymbol{w}_k$:

$$\sum_{k=1}^{K} \|\boldsymbol{w}_k\|_2^2 \leq P. \tag{4.2}$$

Treating interference from other users as noise, for a given tuple of precoders $\boldsymbol{w} = (\boldsymbol{w}_1, \ldots, \boldsymbol{w}_K)$, an achievable rate vector $\boldsymbol{R} = (R_1, \ldots, R_K)$ is given by $\boldsymbol{R} \in \mathbb{R}_+^K : R_k < r_k(\boldsymbol{w})$, with

$$r_k(\boldsymbol{w}) = \log_2\left(1 + \frac{|\boldsymbol{h}_k^{\mathrm{H}} \boldsymbol{w}_k|^2}{\sigma^2 + \sum_{q \neq k} |\boldsymbol{h}_k^{\mathrm{H}} \boldsymbol{w}_q|^2}\right). \tag{4.3}$$

Accordingly, one possible description $(\boldsymbol{r}, \mathcal{X}_1)$ of the physical layer is given by

$$\mathcal{X}_1 = \left\{\boldsymbol{w} \in \mathbb{C}^{KN} : \|\boldsymbol{w}\|_2^2 \leq P\right\}, \tag{4.4}$$

$$\boldsymbol{r} : \mathbb{C}^{KN} \to \mathbb{R}_+^K, \boldsymbol{w} \mapsto (r_1(\boldsymbol{w}), \ldots, r_K(\boldsymbol{w})), \tag{4.5}$$

with $r_k(\boldsymbol{w})$ defined in (4.3). The set $\mathcal{X}_1$ as defined in (4.4) is compact and nonempty,[1] and the rate function $\boldsymbol{r}$ defined by (4.5) is continuous. Accordingly, the pair $(\boldsymbol{r}, \mathcal{X}_1)$ fulfills the requirements of Definitions 2.1.1 and 2.2.1.

Except for special cases, the rate function $\boldsymbol{r}$ is nonconcave.[2]

---

[1] Assuming that $P \geq 0$.

[2] The rate function is concave if $\boldsymbol{h}_k = \boldsymbol{0}$ for all but one users. This case is of little interest as it basically corresponds to a single-user system.

4.1.2 Rate Region

According to Sect. 2.3, the rate region associated with a physical layer $(\boldsymbol{r}, \mathcal{X}_1)$ is given by

$$\mathcal{R} = \boldsymbol{r}(\mathcal{X}_1).$$

As discussed in Chap. 3, the rate region plays an important role in solving the utility maximization problem. A desirable property of the rate region is that it is proper, cf. Definition 2.3.1. As shown in Sect. 2.3.1, any physical layer $(\boldsymbol{r}, \mathcal{X}_1)$ can be extended to have a comprehensive rate region, at the cost of increasing the number of parameters. Proposition 4.1.1 shows that in case of the MISO broadcast channel with linear precoding, such an extension is not necessary, as its rate region is already comprehensive. Based on this result, Proposition 4.1.2 gives a simple condition for the rate region to be proper.

Proposition 4.1.1. *The rate region* $\mathcal{R} = \boldsymbol{r}(\mathcal{X}_1)$*, with* $\mathcal{X}_1$ *as defined in* (4.4) *and* $\boldsymbol{r}$ *as defined in* (4.5) *and* (4.3)*, is comprehensive.*

Proof. Let $\boldsymbol{s}' \in \mathcal{R}$. Accordingly, there exists $\boldsymbol{w}'$ such that $\boldsymbol{s}' = \boldsymbol{r}(\boldsymbol{w}')$. Given $\boldsymbol{w}'$, define a function $\boldsymbol{f} : \mathbb{R}_+^K \to \mathbb{C}^{KN}$ as follows:

$$\boldsymbol{f}(\boldsymbol{p}) = \left(\sqrt{p_1} \frac{\boldsymbol{w}_1'}{\|\boldsymbol{w}_1'\|_2}, \dots, \sqrt{p_K} \frac{\boldsymbol{w}_K'}{\|\boldsymbol{w}_K'\|_2} \right).$$

In the following, it is shown that given $\boldsymbol{s}$, with $\boldsymbol{0} \leq \boldsymbol{s} \leq \boldsymbol{s}'$, there exists $\boldsymbol{p} \in \mathbb{R}_+^K$ with $\boldsymbol{1}^{\mathrm{T}}\boldsymbol{p} \leq P$ such that

$$\boldsymbol{r}(\boldsymbol{f}(\boldsymbol{p})) = \boldsymbol{s},$$

which implies that $\boldsymbol{s} \in \mathcal{R}$.

Consider the function $P_{\min} : \mathbb{R}_+^K \to \mathbb{R}_+ \cup \{\infty\}$, with

$$P_{\min}(\boldsymbol{s}) = \inf_{\boldsymbol{p} \in \mathbb{R}_+^K} \boldsymbol{1}^{\mathrm{T}}\boldsymbol{p} \quad \text{s.t.} \quad \boldsymbol{r}(\boldsymbol{f}(\boldsymbol{p})) \geq \boldsymbol{s}.$$

The constraint $\boldsymbol{r}(\boldsymbol{f}(\boldsymbol{p})) \geq \boldsymbol{s}$ can be rewritten as

$$\tilde{\boldsymbol{p}} \geq \boldsymbol{\iota}(\tilde{\boldsymbol{p}}),$$

where $\tilde{\boldsymbol{p}}$ contains the powers of the users with $s_k > 0$ and $\boldsymbol{\iota}$ fulfills the properties of a *standard interference function* according to the definition of Yates [2], cf. Definition A.9.1. Accordingly, if $P_{\min}(\boldsymbol{s}) < \infty$ then the framework from [2] can be used to compute $P_{\min}(\boldsymbol{s})$. Note that $\boldsymbol{s}' \in \mathcal{R}$, thus $P_{\min}(\boldsymbol{s}') \leq P < \infty$. Let $\boldsymbol{0} \leq \boldsymbol{s} \leq \boldsymbol{s}'$. Then

$$\boldsymbol{r}(\boldsymbol{f}(\boldsymbol{p})) \geq \boldsymbol{s}' \implies \boldsymbol{r}(\boldsymbol{f}(\boldsymbol{p})) \geq \boldsymbol{s},$$

thus $P_{\min}(\boldsymbol{s}) \leq P_{\min}(\boldsymbol{s}') < \infty$. Moreover, by Lemma A.9.2, there exists $\boldsymbol{p}^* \in \mathbb{R}_+^K$ such that

$$\boldsymbol{r}(\boldsymbol{f}(\boldsymbol{p}^*)) = \boldsymbol{s},$$
$$\mathbf{1}^{\mathrm{T}}\boldsymbol{p}^* = P_{\min}(\boldsymbol{s}).$$

It follows that $\mathbf{1}^{\mathrm{T}}\boldsymbol{p}^* \leq P$, thus $\boldsymbol{f}(\boldsymbol{p}^*) \in \mathcal{X}_1$ and $\boldsymbol{s} = \boldsymbol{r}(\boldsymbol{f}(\boldsymbol{p}^*)) \in \mathcal{R}$. □

Proposition 4.1.2. *The rate region $\mathcal{R} = \boldsymbol{r}(\mathcal{X}_1)$, with $\mathcal{X}_1$ as defined in* (4.4) *and $\boldsymbol{r}$ as defined in* (4.5)*, is proper if and only if $P > 0$ and*

$$\boldsymbol{h}_k \neq \mathbf{0}, \forall k.$$

Proof. $\mathcal{R}$ is the image of a compact set under a continuous function, thus $\mathcal{R}$ is compact. By Proposition 4.1.1, $\mathcal{R}$ is comprehensive. Assume that $P > 0$ and $\boldsymbol{h}_k \neq \mathbf{0}, \forall k$. Set

$$\boldsymbol{w}_k = \sqrt{\frac{P}{K}} \frac{\boldsymbol{h}_k}{\|\boldsymbol{h}_k\|_2}, k \in \{1, \ldots, K\}.$$

Clearly, $(\boldsymbol{w}_1, \ldots, \boldsymbol{w}_K) \in \mathcal{X}_1$. Moreover,

$$r_k((\boldsymbol{w}_1, \ldots, \boldsymbol{w}_K)) = \log_2\left(1 + \frac{\frac{P}{K}\|\boldsymbol{h}_k\|_2^2}{\sigma^2 + \sum_{q \neq k} \frac{P}{K}\left|\frac{\boldsymbol{h}_k^{\mathrm{H}}\boldsymbol{h}_q}{\|\boldsymbol{h}_q\|_2}\right|^2}\right) > 0.$$

Thus, there exists $\boldsymbol{s} \in \mathcal{R}$ with $\boldsymbol{s} \gg \mathbf{0}$. In conjunction with the property that $\mathcal{R}$ is comprehensive, this implies that $\mathcal{R}$ has nonempty interior. On the other hand, if $\boldsymbol{h}_k = \mathbf{0}$ for at least one k or $P = 0$, then $r_k(\boldsymbol{w}) = 0, \forall \boldsymbol{w} \in \mathcal{X}_1$ for at least one k, implying that

$$\boldsymbol{r}(\boldsymbol{w}) \in \operatorname{bd}(\mathcal{R}), \forall \boldsymbol{w} \in \mathcal{X}_1,$$

which in turn implies $\operatorname{int}(\mathcal{R}) = \emptyset$. □

Figure 4.1 shows an exemplary rate region of a MISO broadcast channel with linear precoding, for the case of two receivers.[3] Evidently, the rate region is not

[3] The following parameters were used to generate the figure: $P/\sigma^2 = 10$, $\boldsymbol{h}_1 = (2, 1)$, $\boldsymbol{h}_2 = (1, 2)$.

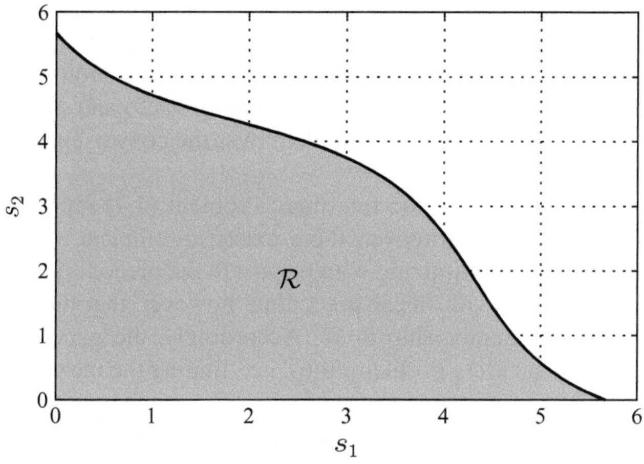

Fig. 4.1 Rate Region of a MISO BC with Linear Precoding, $K = 2$

convex in general. Recall from Sect. 2.3.5 that a convex rate region can be obtained by introducing a time sharing mode.

4.1.3 Intersection Problem

In general, utility maximization in the MISO BC with linear precoding is a nonconvex optimization problem, due to the nonconcavity of the rate map $\boldsymbol{r}$, and it is generally hard to find practically feasible algorithms that converge to the global optimum. There is one problem, however, that can be solved efficiently: the intersection problem

$$\max_{\beta} \beta \quad \text{s.t.} \quad (\boldsymbol{s}' + \beta \boldsymbol{d})^{+} \in \mathcal{R}, \tag{4.6}$$

with $\boldsymbol{s}' \in \mathbb{R}_{+}^{K}$ and $\boldsymbol{d} \in \mathbb{R}_{++}^{K}$, cf. Sect. 2.3.2. In the remainder of this section, the parameterization of the physical layer from Sect. 4.1.1 is used. For clarity, the set of feasible physical layer parameters is denoted by

$$\mathcal{W} = \left\{ \boldsymbol{w} \in \mathbb{C}^{KN} : \|\boldsymbol{w}\|_2^2 \leq P \right\},$$

instead of the general notation $\mathcal{X}_1$.

The intersection problem in parameter space formulation is given by

$$\max_{(\beta, \boldsymbol{w}) \in \mathbb{R} \times \mathcal{W}} \beta \quad \text{s.t.} \quad (\boldsymbol{s}' + \beta \boldsymbol{d})^{+} \leq \boldsymbol{r}(\boldsymbol{w}). \tag{4.7}$$

Observe that the parameters of Problem (4.7) are the precoders only, there is no time sharing mode provided. The difference between providing a time sharing mode or not is significant. As shown in the following, the intersection problem over the rate region $\mathcal{R} = \boldsymbol{r}(\mathcal{X}_1)$, with $\boldsymbol{r}$ and $\mathcal{X}_1$ as defined in (4.5) and (4.4), is efficiently solvable. In contrast, the intersection problem over the convex hull of $\mathcal{R}$ represents a hard problem.

Due to the nonconcavity of the rate map, Problem (4.7) represents a nonconvex optimization problem. Moreover, there exists no efficient algorithm to solve Problem (4.7) directly by optimizing with respect to the precoders $\boldsymbol{w}$. It is a specific feature of the MISO BC with linear precoding, however, that there exist efficient algorithms to test for membership in $\mathcal{R}$. Accordingly, the generalized max–min problem is solved by tackling Problem (4.6), i.e., finding the maximum β such that $(\boldsymbol{s}' + \beta \boldsymbol{d})^+ \in \mathcal{R}$.

To this end, first define an SINR function $\boldsymbol{\gamma} : \mathbb{C}^{KN} \to \mathbb{R}_+^K, \boldsymbol{w} \mapsto (\gamma_1(\boldsymbol{w}), \ldots, \gamma_K(\boldsymbol{w}))$, with

$$\gamma_k(\boldsymbol{w}) = \frac{|\boldsymbol{h}_k^{\mathrm{H}} \boldsymbol{w}_k|^2}{\sigma^2 + \sum_{q \neq k} |\boldsymbol{h}_k^{\mathrm{H}} \boldsymbol{w}_q|^2}.$$

The SINR function $\boldsymbol{\gamma}$ and the rate function $\boldsymbol{r}$ defined in (4.5) are closely related:

$$\boldsymbol{r} = \boldsymbol{\rho} \circ \boldsymbol{\gamma},$$

where $\boldsymbol{\rho}$ is bijective and increasing, with

$$\rho_k(\boldsymbol{y}) = \log_2(1 + y_k).$$

Likewise, the inverse function $\boldsymbol{\rho}^{-1}$, with

$$\rho_k^{-1}(\boldsymbol{s}) = 2^{s_k} - 1,$$

is also increasing. A rate $\boldsymbol{s} \in \mathbb{R}_+^K$ is an element of the comprehensive rate region $\mathcal{R}$ if there exists $\boldsymbol{w} \in \mathcal{W}$ such that

$$\boldsymbol{s} \leq \boldsymbol{r}(\boldsymbol{w}). \tag{4.8}$$

Using the fact that $\boldsymbol{\rho}^{-1}$ is bijective and increasing, Inequality (4.8) is equivalent to

$$\boldsymbol{\rho}^{-1}(\boldsymbol{s}) \leq \boldsymbol{\rho}^{-1}(\boldsymbol{r}(\boldsymbol{w})) = \boldsymbol{\gamma}(\boldsymbol{w}). \tag{4.9}$$

Define an SINR balancing function $C : \mathbb{R}_+^K \setminus \{\boldsymbol{0}\} \to \mathbb{R}$ as follows:

$$C(\bar{\boldsymbol{\gamma}}) = \max_{\boldsymbol{w} \in \mathcal{W}} \min_{k \in \mathcal{I}(\bar{\boldsymbol{\gamma}})} \frac{\gamma_k(\boldsymbol{w})}{\bar{\gamma}_k}, \tag{4.10}$$

where

$$\mathcal{I}(\boldsymbol{x}) = \{k \in \{1, \ldots, K\} : x_k > 0\}.$$

The SINR function $\boldsymbol{\gamma}$ is continuous, and $\mathcal{W}$ is compact, thus the right-hand side of (4.10) is well-defined for $\bar{\boldsymbol{\gamma}} \in \mathbb{R}_+^K \setminus \{\mathbf{0}\}$, and the maximum is attained in $\mathcal{W}$. Most importantly, the SINR balancing problem (4.10) is efficiently solvable [73].

According to Inequality (4.9) and the definition of C, a rate vector $\boldsymbol{s} \in \mathbb{R}_+^K \setminus \{\mathbf{0}\}$ is element of $\mathcal{R}$ if and only if

$$C(\rho^{-1}(\boldsymbol{s})) \geq 1. \tag{4.11}$$

As a result, the SINR balancing function provides a membership test for $\mathcal{R}$ that is efficiently computable, using the algorithm from [73]. In fact, the SINR balancing problem can be used to test for membership in $\mathcal{R} \setminus \{\mathbf{0}\}$. Moreover, $\mathbf{0} \in \mathcal{R}$ for any proper rate region.

The intersection problem is solved by combining the membership test in $\mathcal{R}$ with a bisection procedure. Define $\phi : \mathbb{R} \to \mathbb{R}^K$,

$$\phi(\beta) = (\boldsymbol{s}' + \beta \boldsymbol{d})^+,$$

with $\boldsymbol{s}' \in \mathbb{R}_+^K$ and $\boldsymbol{d} \in \mathbb{R}_{++}^K$, and let

$$\beta_0 = -\max_k \frac{s'_k}{d_k}. \tag{4.12}$$

Then $\phi(\beta_0) = \mathbf{0}$. Define a function $f : \mathbb{R} \to \mathbb{R}$,

$$f(\beta) = \begin{cases} C(\rho^{-1}(\phi(\beta))) - 1, & \beta > \beta_0, \\ 1, & \beta \leq \beta_0. \end{cases}$$

The function ϕ is strictly increasing on $[\beta_0, \infty)$, hence

$$\phi(\beta) > \mathbf{0}, \forall \beta > \beta_0, \tag{4.13}$$

showing that f is well-defined. The function f is introduced to circumvent the problem that C is not defined at $\mathbf{0}$. From $\mathcal{R}$ proper it follows that $\mathbf{0} \in \mathcal{R}$, hence setting $f(\beta)$ to an arbitrary positive value for $\beta \leq \beta_0$ yields

$$f(\beta) \geq 0 \iff \phi(\beta) \in \mathcal{R}. \tag{4.14}$$

Moreover,

$$\phi(\beta') \in \mathcal{R}, \beta' \geq \beta \implies \mathbf{0} \leq \phi(\beta) \leq \phi(\beta') \implies \phi(\beta) \in \mathcal{R}. \tag{4.15}$$

Let

$$\beta^* = \max_{\beta \in \mathbb{R}} \beta \quad \text{s.t.} \quad \phi(\beta) \in \mathcal{R}.$$

By (4.14) and (4.15),

$$f(\beta) \geq 0 \iff \beta \leq \beta^*.$$

As a result, β^* can be computed by bisection on f. Let $\underline{\beta}^{(n)}$ and $\overline{\beta}^{(n)}$ denote upper and lower bound on β^* in the bisection procedure, respectively. The purpose of computing β^* is to find the intersection point $\phi(\beta^*)$ in wpar ($\mathcal{R}$), while the value of f around β^* is of little significance. Consequently, the distance in rate space is used as convergence criterion for the bisection:

$$\epsilon^{(n)} = \left\| \phi(\overline{\beta}^{(n)}) - \phi(\underline{\beta}^{(n)}) \right\|_2.$$

The rate vector $\phi(\underline{\beta}^{(n)})$ is a feasible solution. Moreover, from

$$\phi(\underline{\beta}^{(n)}) \leq \phi(\beta^*) \leq \phi(\overline{\beta}^{(n)})$$

it follows that

$$\epsilon^{(n)} \geq \left\| \phi(\beta^*) - \phi(\underline{\beta}^{(n)}) \right\|_2.$$

Finally, by continuity of ϕ, $\epsilon^{(n)}$ goes to zero as $\overline{\beta}^{(n)} - \underline{\beta}^{(n)}$ goes to zero.[4]

The bisection procedure requires initial values for the bounds $\underline{\beta}^{(n)}$ and $\overline{\beta}^{(n)}$. For the lower bound, choose $\underline{\beta}^{(0)} = \beta_0$. An upper bound follows from the utopia point $\boldsymbol{s}^{\mathrm{U}}$:

$$\phi(\beta) \leq \boldsymbol{s}^{\mathrm{U}} \implies \beta \leq \min_k \frac{s_k^{\mathrm{U}} - s_k'}{d_k}.$$

[4]Another benefit of using the distance in rate space as convergence criterion is that an examination of the continuity properties of f is not required. The function f is obviously not continuous at β_0, and continuity at $\beta > \beta_0$ is not obvious.

4.1.4 An Alternative Parameterization

The parameterization of the physical layer in terms of the precoding vectors $(\boldsymbol{w}_1, \ldots, \boldsymbol{w}_K)$ is not the only possible parameterization. Using the duality relation between uplink and downlink [73], the uplink powers $\boldsymbol{p} \in \mathbb{R}_+^K$ can be used to parameterize the physical layer. In this case, the description $(\boldsymbol{r}, \mathcal{X}_1)$ of the physical layer is given by

$$\mathcal{X}_1 = \left\{\boldsymbol{p} \in \mathbb{R}_+^K : \mathbf{1}^{\mathrm{T}}\boldsymbol{p} \leq P\right\}, \tag{4.16}$$

$$\boldsymbol{r} : \mathbb{R}_+^K \to \mathbb{R}_+^K, \boldsymbol{p} \mapsto (r_1(\boldsymbol{p}), \ldots, r_K(\boldsymbol{p})), \tag{4.17}$$

with

$$r_k(\boldsymbol{p}) = \log_2 \det\left(\boldsymbol{I} + \boldsymbol{B}_k(\boldsymbol{p})^{-1}\boldsymbol{h}_k\boldsymbol{h}_k^{\mathrm{H}} p_k\right) \tag{4.18}$$

$$= \log_2(1 + p_k \boldsymbol{h}_k^{\mathrm{H}}\boldsymbol{B}_k(\boldsymbol{p})^{-1}\boldsymbol{h}_k), \tag{4.19}$$

and

$$\boldsymbol{B}_k(\boldsymbol{p}) = \sigma^2\boldsymbol{I} + \sum_{q \neq k} \boldsymbol{h}_q\boldsymbol{h}_q^{\mathrm{H}} p_q.$$

As shown in [73], both parameterizations are equivalent in the sense that their rate regions are identical. The rate functions of both parameterizations are nonconcave.

4.1.5 Global Parameterization of the Pareto Set

The parameterization of the rate region by uplink powers introduced in Sect. 4.1.4 is useful for characterizing the Pareto boundary of $\mathcal{R}$.

For $k \in \{1, \ldots, K\}$, define a function $f_k : \mathbb{R}_+^K \to \mathbb{R}$ as follows:

$$f_k(\boldsymbol{p}) = \boldsymbol{h}_k^{\mathrm{H}}\boldsymbol{B}_k(\boldsymbol{p})^{-1}\boldsymbol{h}_k.$$

Lemma 4.1.3. *The function f_k is decreasing. Moreover, if $\mathcal{R}$ is proper, then*

$$f_k(\boldsymbol{p}) > 0, \forall \boldsymbol{p} \in \mathbb{R}_+^K, \tag{4.20}$$

and

$$\alpha > 1 \implies f_k(\alpha\boldsymbol{p}) > \alpha^{-1} f_k(\boldsymbol{p}).$$

Proof. Consider the directional derivative of f_k at $\boldsymbol{p} \in \mathbb{R}_+^K$ in direction $\boldsymbol{u} \in \mathbb{R}_+^K$:

$$f_k'(\boldsymbol{p}, \boldsymbol{u}) = \left.\frac{\partial f_k(\boldsymbol{p} + t\boldsymbol{u})}{\partial t}\right|_{t=0} = -\boldsymbol{h}_k^{\mathrm{H}} \boldsymbol{B}_k(\boldsymbol{p})^{-1} \boldsymbol{A}_k(\boldsymbol{u}) \boldsymbol{B}_k(\boldsymbol{p})^{-1} \boldsymbol{h}_k,$$

with

$$\boldsymbol{A}_k(\boldsymbol{p}) = \sum_{q \neq k} \boldsymbol{h}_q \boldsymbol{h}_q^{\mathrm{H}} p_q.$$

The matrix $\boldsymbol{A}_k(\boldsymbol{u})$ is positive semidefinite for all $\boldsymbol{u} \in \mathbb{R}_+^K$, and $\boldsymbol{B}_k(\boldsymbol{p})^{-1}$ is Hermitian, hence

$$f_k'(\boldsymbol{p}, \boldsymbol{u}) \leq 0, \forall (\boldsymbol{p}, \boldsymbol{u}) \in \mathbb{R}_+^K \times \mathbb{R}_+^K,$$

showing that f_k is decreasing.

Let R denote the rank of $\boldsymbol{A}_k(\boldsymbol{p})$. Moreover, let β_r and $\boldsymbol{v}_r$ denote the r-th eigenvalue and eigenvector of $\boldsymbol{A}_k(\boldsymbol{p})$, respectively. Assume that the eigenvalues are sorted in descending order. It follows that the r-th eigenvalue of $\boldsymbol{A}_k(\alpha \boldsymbol{p})$, with $\alpha > 1$, is $\alpha \beta_r$, while the eigenvectors remain unchanged under scaling. The eigenvalue decomposition of $\boldsymbol{B}_k(\alpha \boldsymbol{p})$ is given by

$$\boldsymbol{B}_k(\alpha \boldsymbol{p}) = \sigma^2 \boldsymbol{I} + \boldsymbol{A}_k(\alpha \boldsymbol{p}) = \sum_{r=1}^{R} (\sigma^2 + \alpha \beta_r) \boldsymbol{v}_r \boldsymbol{v}_r^{\mathrm{H}} + \sum_{r=R+1}^{N} \sigma^2 \boldsymbol{v}_r \boldsymbol{v}_r^{\mathrm{H}},$$

where, in case of $R < N$, $\{\boldsymbol{v}_{R+1}, \ldots, \boldsymbol{v}_N\}$ is an orthonormal basis of null $(\boldsymbol{A}_k(\alpha \boldsymbol{p}))$. Hence,

$$f_k(\alpha \boldsymbol{p}) = \sum_{r=1}^{R} (\sigma^2 + \alpha \beta_r)^{-1} |\boldsymbol{h}_k^{\mathrm{H}} \boldsymbol{v}_r|^2 + \sum_{r=R+1}^{N} \sigma^{-2} |\boldsymbol{h}_k^{\mathrm{H}} \boldsymbol{v}_r|^2.$$

For $\alpha > 1$,

$$\sigma^{-2} > \alpha^{-1} \sigma^{-2},$$
$$(\sigma^2 + \alpha \beta_r)^{-1} > \alpha^{-1} (\sigma^2 + \beta_r)^{-1}.$$

By Proposition 4.1.2, $\boldsymbol{h}_k \neq \boldsymbol{0}$. Moreover, span $(\{\boldsymbol{v}_1, \ldots, \boldsymbol{v}_N\}) = \mathbb{C}^N$. Hence, there exists $r \in \{1, \ldots, N\}$ such that

$$\boldsymbol{h}_k^{\mathrm{H}} \boldsymbol{v}_r \neq \boldsymbol{0}.$$

As a result,

$$f_k(\boldsymbol{p}) > 0, \forall \boldsymbol{p} \in \mathbb{R}_+^K,$$

and

$$
\begin{aligned}
f_k(\alpha \boldsymbol{p}) &= \sum_{r=1}^{R} (\sigma^2 + \alpha\beta_r)^{-1} |\boldsymbol{h}_k^{\mathrm{H}} \boldsymbol{v}_r|^2 + \sum_{r=R+1}^{N} \sigma^{-2} |\boldsymbol{h}_k^{\mathrm{H}} \boldsymbol{v}_r|^2 \\
&> \sum_{r=1}^{R} \alpha^{-1} (\sigma^2 + \beta_r)^{-1} |\boldsymbol{h}_k^{\mathrm{H}} \boldsymbol{v}_r|^2 + \sum_{r=R+1}^{N} \alpha^{-1} \sigma^{-2} |\boldsymbol{h}_k^{\mathrm{H}} \boldsymbol{v}_r|^2 \\
&= \alpha^{-1} f_k(\boldsymbol{p}).
\end{aligned}
$$

□

Lemma 4.1.4. *Let $\mathcal{R}$ be proper and $\boldsymbol{p} \in \mathbb{R}_+^K$. A necessary condition for $\boldsymbol{r}(\boldsymbol{p})$ to be an element of* wpar ($\mathcal{R}$) *is that the uplink power allocation $\boldsymbol{p}$ uses all available power, i.e.,*

$$
\boldsymbol{r}(\boldsymbol{p}) \in \operatorname{wpar}(\mathcal{R}) \implies \mathbf{1}^{\mathrm{T}} \boldsymbol{p} = P.
$$

Proof. Note that $\boldsymbol{r}(\mathbf{0}) \notin \operatorname{wpar}(\mathcal{R})$, due to the fact that $\mathcal{R}$ is proper. Thus, it is sufficient to consider $\boldsymbol{p} \in \mathbb{R}_+^K \setminus \{\mathbf{0}\}$. Let $\boldsymbol{p} \in \mathbb{R}_+^K \setminus \{\mathbf{0}\}$ such that $\mathbf{1}^{\mathrm{T}} \boldsymbol{p} < P$. It follows that $\boldsymbol{r}(\boldsymbol{p}) \in \mathcal{R}$. Define

$$
\alpha = \frac{1}{2} \left(1 + \frac{P}{\mathbf{1}^{\mathrm{T}} \boldsymbol{p}} \right).
$$

Then

$$
\alpha \mathbf{1}^{\mathrm{T}} \boldsymbol{p} = \frac{1}{2} (\mathbf{1}^{\mathrm{T}} \boldsymbol{p} + P) < P,
$$

thus $\boldsymbol{r}(\alpha \boldsymbol{p}) \in \mathcal{R}$. Noting that $\alpha > 1$, by Lemma 4.1.3,

$$
\alpha p_k f_k(\alpha \boldsymbol{p}) > \alpha p_k \alpha^{-1} f_k(\boldsymbol{p}) = p_k f_k(\boldsymbol{p}), \quad \forall k : p_k > 0,
$$

implying that

$$
r_k(\alpha \boldsymbol{p}) > r_k(\boldsymbol{p}), \quad \forall k : p_k > 0. \tag{4.21}
$$

If $p_k > 0$ for all k, it immediately follows that $\boldsymbol{r}(\boldsymbol{p})$ is not weak Pareto optimal. If $\boldsymbol{p}$ has zero components, additional steps are necessary to show existence of a point $\boldsymbol{s} \in \mathcal{R}$ with $\boldsymbol{s} \gg \boldsymbol{r}(\boldsymbol{p})$. First, observe that by (4.20),

$$
p_k' > 0 \implies r_k(\boldsymbol{p}') > 0. \tag{4.22}
$$

Let

$$
\boldsymbol{p}'(\eta) = \alpha \boldsymbol{p} + \eta \mathbf{1},
$$

and

$$\overline{\eta} = \frac{1}{2K}(P - \mathbf{1}^{\mathrm{T}}\boldsymbol{p}).$$

Observe that $\overline{\eta} > 0$ and

$$\boldsymbol{r}(\boldsymbol{p}'(\eta)) \in \mathcal{R}, \forall \eta \in [0, \overline{\eta}]. \tag{4.23}$$

Moreover, by (4.21) and continuity of $\boldsymbol{r}$, there exists $\eta \in (0, \overline{\eta})$ such that

$$r_k(\boldsymbol{p}'(\eta)) > r_k(\boldsymbol{p}), \forall k : p_k > 0. \tag{4.24}$$

Equations (4.22), (4.23) and (4.24) imply that there exists $\eta > 0$ such that $\boldsymbol{r}(\boldsymbol{p}'(\eta)) \in \mathcal{R}$ and

$$\boldsymbol{r}(\boldsymbol{p}'(\eta)) \gg \boldsymbol{r}(\boldsymbol{p}),$$

hence $\boldsymbol{r}(\boldsymbol{p})$ is not weak Pareto optimal. □

Define a function $\boldsymbol{\iota} : \mathbb{R}_+^K \times \mathbb{R}_{++}^K \to \mathbb{R}^K$ as follows:

$$\iota_k(\boldsymbol{p}, \boldsymbol{s}) = \frac{2^{s_k} - 1}{f_k(\boldsymbol{p})}.$$

Lemma 4.1.5. *Let $\mathcal{R}$ be proper and $\boldsymbol{s} \in \mathbb{R}_{++}^K$. Then $\boldsymbol{\iota}(\cdot, \boldsymbol{s})$ is a standard interference function according to Definition A.9.1.*

Proof. By Lemma 4.1.3 and $\boldsymbol{s} \in \mathbb{R}_{++}^K$,

$$\begin{aligned}
&\boldsymbol{\iota}(\boldsymbol{p}, \boldsymbol{s}) \gg \mathbf{0}, \\
\boldsymbol{p} \geq \boldsymbol{p}' &\implies \boldsymbol{\iota}(\boldsymbol{p}, \boldsymbol{s}) \geq \boldsymbol{\iota}(\boldsymbol{p}', \boldsymbol{s}), \\
\alpha > 1 &\implies \boldsymbol{\iota}(\alpha \boldsymbol{p}, \boldsymbol{s}) < \alpha \boldsymbol{\iota}(\boldsymbol{p}, \boldsymbol{s}).
\end{aligned}$$

□

Lemma 4.1.6. *Let $\boldsymbol{p}, \boldsymbol{p}' \in \mathbb{R}_{++}^K$. Then*

$$\boldsymbol{r}(\boldsymbol{p}') > \boldsymbol{r}(\boldsymbol{p}) \implies \boldsymbol{p}' > \boldsymbol{p}.$$

Proof. From $\boldsymbol{p}, \boldsymbol{p}' \in \mathbb{R}_{++}^K$ it follows that $\boldsymbol{r}(\boldsymbol{p}) \in \mathbb{R}_{++}^K$ and $\boldsymbol{r}(\boldsymbol{p}') \in \mathbb{R}_{++}^K$. Let $\boldsymbol{r}(\boldsymbol{p}') > \boldsymbol{r}(\boldsymbol{p})$. Then

$$p'_k \geq \frac{p_k f_k(\boldsymbol{p})}{f_k(\boldsymbol{p}')} = \iota_k(\boldsymbol{p}', \boldsymbol{r}(\boldsymbol{p})), \forall k.$$

Moreover, there exists a k such that the inequality is strict. Thus,

$$\boldsymbol{p}' > \iota(\boldsymbol{p}', \boldsymbol{r}(\boldsymbol{p})).$$

By definition of ι,

$$\boldsymbol{p} = \iota(\boldsymbol{p}, \boldsymbol{r}(\boldsymbol{p})).$$

Hence, by Corollary A.9.3,

$$\boldsymbol{p}' > \iota(\boldsymbol{p}', \boldsymbol{r}(\boldsymbol{p})) \implies \boldsymbol{p}' > \boldsymbol{p}. \tag{4.25}$$

□

Proposition 4.1.7. *Let $\mathcal{R}$ be proper and $\boldsymbol{p} \in \mathbb{R}^K_{++}$. Then*

$$\mathbf{1}^{\mathrm{T}}\boldsymbol{p} = P \implies \boldsymbol{r}(\boldsymbol{p}) \in \operatorname{par}(\mathcal{R}).$$

Proof. Assume that $\mathbf{1}^{\mathrm{T}}\boldsymbol{p} = P$. From $\mathcal{R}$ proper and $\boldsymbol{p} \in \mathbb{R}^K_{++}$ it follows that $\boldsymbol{r}(\boldsymbol{p}) \in \mathbb{R}^K_{++}$. Assume there exists $\boldsymbol{p}' \in \mathbb{R}^K_{++}$ such that $\boldsymbol{r}(\boldsymbol{p}') > \boldsymbol{r}(\boldsymbol{p})$. By Lemma 4.1.6,

$$\mathbf{1}^{\mathrm{T}}\boldsymbol{p}' > \mathbf{1}^{\mathrm{T}}\boldsymbol{p} = P.$$

Hence, $\boldsymbol{r}(\boldsymbol{p}') \notin \mathcal{R}$, showing that $\boldsymbol{r}(\boldsymbol{p})$ is Pareto optimal. □

Corollary 4.1.8.

$$\operatorname{wpar}(\mathcal{R}) \cap \mathbb{R}^K_{++} = \operatorname{par}(\mathcal{R}) \cap \mathbb{R}^K_{++}.$$

Lemma 4.1.6, Proposition 4.1.7 and Corollary 4.1.8 only apply to rate vectors for which all link rates are strictly positive. In order to obtain a characterization of the complete Pareto set, a consideration of points with zero link rates is necessary. As before, links with zero link rate are denoted as deactivated links. As in Sect. 3.5.5, let $\mathcal{D}$ contain the indices of the deactivated links and define a set $\mathcal{Z}_{\mathcal{D}}$ as follows:

$$\mathcal{Z}_{\mathcal{D}} = \left\{\boldsymbol{x} \in \mathbb{R}^K_+ : x_k = 0, \forall k \in \mathcal{D}\right\} \cap \left\{\boldsymbol{x} \in \mathbb{R}^K_+ : x_k > 0, \forall k \notin \mathcal{D}\right\}.$$

Under the assumption that $\mathcal{R}$ is proper, $r_k(\boldsymbol{p}) = 0$ is equivalent to $p_k = 0$:

$$p \in \mathcal{Z}_\mathcal{D} \iff r(p) \in \mathcal{Z}_\mathcal{D}.$$

In other words, a link that is deactivated is assigned zero power.

Evidently, by removing the deactivated links from the description, a system with $K - \text{card}(\mathcal{D})$ links results. The remaining links are all active, which immediately yields the following corollaries:

Corollary 4.1.9. *Let* $p, p' \in \mathcal{Z}_\mathcal{D}$. *Then*

$$r(p') > r(p) \implies p' > p.$$

Corollary 4.1.10. *Let* $\mathcal{R}$ *be proper and* $p \in \mathcal{Z}_\mathcal{D}$. *Then*

$$\mathbf{1}^\mathrm{T} p = P \implies r(p) \in \text{par}(\mathcal{R} \cap \mathcal{Z}_\mathcal{D}). \tag{4.26}$$

From Corollary 4.1.10 it can be conjectured that

$$\mathbf{1}^\mathrm{T} p = P \implies r(p) \in \text{par}(\mathcal{R}),$$

for *any* $p \in \mathbb{R}_+^K$, as (4.26) holds for any set of deactivated links. However, (4.26) only states that a rate vector s in $\mathcal{Z}_\mathcal{D}$ which uses full power is not dominated by any other vector in $\mathcal{R} \cap \mathcal{Z}_\mathcal{D}$. There could still exist a vector in $\mathcal{R} \cap \mathcal{Z}_{\mathcal{D}'}$, with $\mathcal{D}' \subset \mathcal{D}$, that dominates s, which would imply $s \notin \text{par}(\mathcal{R})$. Such a scenario is illustrated in Fig. 4.2. In Fig. 4.2, the point s is Pareto optimal in $\mathcal{R} \cap \mathcal{Z}_{\{2\}}$, but is dominated in $\mathcal{R}$ by s'. However, Fig. 4.2 already indicates that such a point s' does not exist in the rate region of the MISO BC with linear precoding. In Fig. 4.2 there are multiple

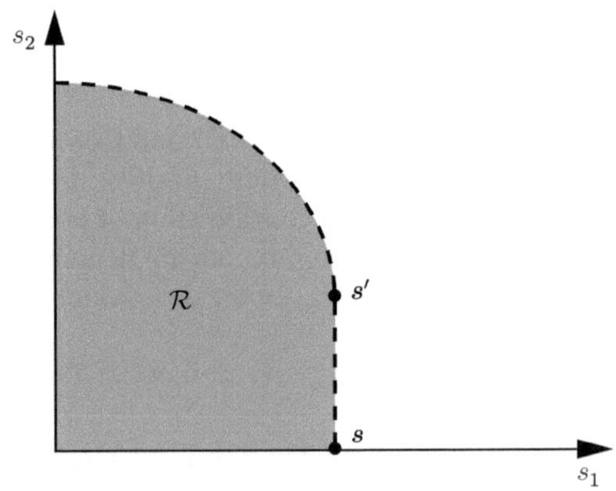

Fig. 4.2 Example of a point s that is Pareto optimal in $\mathcal{R} \cap \mathcal{Z}_{\{2\}}$, but not in $\mathcal{R}$

weak Pareto optimal points in $\mathbb{R}^2_{++}$ that are not Pareto optimal, which contradicts Corollary 4.1.8. The argument is made strict in the following proposition.

Proposition 4.1.11. *Let* $\boldsymbol{s} \in \operatorname{par}(\mathcal{R} \cap \mathcal{Z}_{\mathcal{D}})$, *with* $\mathcal{R}$ *proper. Then* $\boldsymbol{s}$ *is Pareto optimal in* $\mathcal{R}$, *i.e.,*

$$\operatorname{par}(\mathcal{R} \cap \mathcal{Z}_{\mathcal{D}}) \subseteq \operatorname{par}(\mathcal{R}).$$

Proof. Choose $\boldsymbol{p} \in \mathcal{Z}_{\mathcal{D}}$ such that $\boldsymbol{r}(\boldsymbol{p}) \in \operatorname{par}(\mathcal{R} \cap \mathcal{Z}_{\mathcal{D}})$. By Proposition A.4.6, $\boldsymbol{r}(\boldsymbol{p}) \in \operatorname{wpar}(\mathcal{R})$. Hence, by Lemma 4.1.4, $\mathbf{1}^{\mathrm{T}}\boldsymbol{p} = P$. Assume there exists $\boldsymbol{p}' \in \mathcal{Z}_{\mathcal{D}'}$ such that $\boldsymbol{r}(\boldsymbol{p}') \in \mathcal{R}$ and

$$\boldsymbol{r}(\boldsymbol{p}') > \boldsymbol{r}(\boldsymbol{p}). \tag{4.27}$$

Equation (4.27) in conjunction with the Pareto optimality of $\boldsymbol{r}(\boldsymbol{p})$ in $\mathcal{R} \cap \mathcal{Z}_{\mathcal{D}}$ implies $\mathcal{D}' \subset \mathcal{D}$ (i.e., $\boldsymbol{r}(\boldsymbol{p}')$ has less zero entries than $\boldsymbol{r}(\boldsymbol{p})$, cf. Fig. 4.2). Hence, Corollary 4.1.9 cannot be applied directly.[5] Let

$$\overline{\boldsymbol{s}} = \frac{1}{2}(\boldsymbol{r}(\boldsymbol{p}') + \boldsymbol{r}(\boldsymbol{p})).$$

Clearly, $\boldsymbol{r}(\boldsymbol{p}) < \overline{\boldsymbol{s}} < \boldsymbol{r}(\boldsymbol{p}')$ and $\overline{\boldsymbol{s}} \in \mathcal{Z}_{\mathcal{D}'}$. By comprehensiveness of $\mathcal{R}$, $\boldsymbol{r}(\boldsymbol{p}') \in \mathcal{R}$ implies $\overline{\boldsymbol{s}} \in \mathcal{R}$. Moreover,

$$\boldsymbol{r}(\boldsymbol{p}) \in \operatorname{wpar}(\mathcal{R}), \overline{\boldsymbol{s}} \in \mathcal{R}, \boldsymbol{r}(\boldsymbol{p}) < \overline{\boldsymbol{s}} \implies \overline{\boldsymbol{s}} \in \operatorname{wpar}(\mathcal{R}).$$

Thus, by Lemma 4.1.4, $\overline{\boldsymbol{s}} = \boldsymbol{r}(\overline{\boldsymbol{p}})$ with $\mathbf{1}^{\mathrm{T}}\overline{\boldsymbol{p}} = P$. As both $\overline{\boldsymbol{p}}$ and $\boldsymbol{p}'$ are element of $\mathcal{Z}_{\mathcal{D}'}$, Corollary 4.1.9 can be applied, yielding $\mathbf{1}^{\mathrm{T}}\boldsymbol{p}' > P$, which contradicts $\boldsymbol{r}(\boldsymbol{p}') \in \mathcal{R}$. □

Corollary 4.1.12. *Let* $\mathcal{R}$ *be proper. Then*

$$\operatorname{wpar}(\mathcal{R}) = \operatorname{par}(\mathcal{R}).$$

By Corollary 4.1.12 and Proposition 2.3.7, the set $\operatorname{par}(\mathcal{R}) \cap \mathbb{R}^K_{++}$ is a $K-1$ dimensional manifold. By a continuous extension, it is possible to find a manifold $\mathcal{M}$ such that $\mathcal{M}$ has a global parameterization in closed-form and

$$\mathcal{M} \cap \mathbb{R}^K_{+} = \operatorname{par}(\mathcal{R}),$$

[5] Corollary 4.1.9 requires $\mathcal{D} = \mathcal{D}'$.

i.e., the global parameterization of $\mathcal{M}$ covers the *whole* Pareto set. To this end, note that the matrix $\boldsymbol{B}_k(\boldsymbol{p})$ is positive definite on $\mathbb{R}_+^K$ and continuous in $\boldsymbol{p}$. As a result, there exists $\epsilon_k > 0$ such that $\boldsymbol{B}_k(\boldsymbol{p})$ is positive definite for all $\boldsymbol{p} > -\epsilon_k \mathbf{1}$. Similarly, the term

$$a_k(\boldsymbol{p}) = 1 + p_k \boldsymbol{h}_k^{\mathrm{H}} \boldsymbol{B}_k(\boldsymbol{p})^{-1} \boldsymbol{h}_k$$

is positive on $\mathbb{R}_+^K$ and continuous in $\boldsymbol{p}$. Hence, there exists δ_k, with $0 < \delta_k < \epsilon_k$, such that $a_k(\boldsymbol{p}) > 0$ for all $\boldsymbol{p} > -\delta_k \mathbf{1}$. Let

$$\mathcal{S} = \{-\delta \mathbf{1}\} + \mathbb{R}_{++}^K,$$

where

$$\delta = \min_k \delta_k .$$

Define a function $\bar{\boldsymbol{r}} : \mathcal{S} \to \mathbb{R}^K$,

$$\bar{r}_k(\boldsymbol{p}) = \log_2 \left(a_k(\boldsymbol{p})\right), \quad k \in \{1, \ldots, K\} .$$

The function $\bar{\boldsymbol{r}}$ is obtained by extending the domain of $\boldsymbol{r}$. For notational convenience, in the following no distinction will be made between $\bar{\boldsymbol{r}}$ and $\boldsymbol{r}$. If $\boldsymbol{r}$ is evaluated at $\boldsymbol{p} \in \mathcal{S} \setminus \mathbb{R}_+^K$, it is understood that the definition of $\bar{\boldsymbol{r}}$ applies. Let

$$\mathcal{M} = \left\{ \boldsymbol{r}(\boldsymbol{p}) : \boldsymbol{p} \in \mathcal{S}, \mathbf{1}^{\mathrm{T}} \boldsymbol{p} = P \right\} .$$

If $\mathcal{R}$ is proper, $\boldsymbol{p} \in \mathcal{S} \setminus \mathbb{R}_+^K$ implies that there exists k such that $r_k(\boldsymbol{p}) < 0$. As a result,

$$\mathcal{M} \cap \mathbb{R}_+^K = \mathrm{par}\,(\mathcal{R}) .$$

Moreover, by shifting $\mathcal{M}$ by $\delta \mathbf{1}$ and applying similar arguments as in Proposition 2.3.7, it can be shown that $\mathcal{M}$ is a $K - 1$ dimensional manifold, provided that $\mathcal{R}$ is proper. The partial derivative of r_k with respect to power p_q at $\boldsymbol{p}$ is given by

$$\frac{\partial r_k}{\partial p_q}(\boldsymbol{p}) = \begin{cases} \frac{p_k}{\ln(2) a_k(\boldsymbol{p})}, & k = q, \\ -\frac{1}{\ln(2) a_k(\boldsymbol{p})} \boldsymbol{h}_k^{\mathrm{H}} \boldsymbol{B}_k(\boldsymbol{p})^{-1} \boldsymbol{h}_q \boldsymbol{h}_q^{\mathrm{H}} \boldsymbol{B}_k(\boldsymbol{p})^{-1} \boldsymbol{h}_k, & k \neq q. \end{cases}$$

The partial derivatives are continuous on $\mathcal{S}$, hence $\boldsymbol{r}$ is C^1.

Assumption 4.1.1. *The gradient matrix* $\nabla \boldsymbol{r}(\boldsymbol{p}) \in \mathbb{R}^{K \times K}$ *has full rank for all* $\boldsymbol{p} \in \mathcal{S}$.

By Assumption 4.1.1, the restriction of $\boldsymbol{r}$ to $\boldsymbol{r}^{-1}(\mathcal{M})$ is a C^1 diffeomorphism. Based on this result, obtaining a differentiable global parameterization of $\mathcal{M}$ is straightforward. Let $\boldsymbol{Q} \in \mathbb{R}^{K \times K-1}$ be a matrix such that the columns of $\boldsymbol{Q}$ constitute a basis of null $(\mathbf{1}^{\mathrm{T}})$:

$$\operatorname{span}(\boldsymbol{Q}) = \operatorname{null}\left(\mathbf{1}^{\mathrm{T}}\right).$$

Given $\boldsymbol{Q}$, define a set $\mathcal{U}$ and a map $\phi : \mathcal{U} \to \mathcal{S}$ as follows:

$$\mathcal{U} = \left\{\boldsymbol{\mu} \in \mathbb{R}^{K-1} : \frac{P}{K}\mathbf{1} + \boldsymbol{Q}\boldsymbol{\mu} \in \mathcal{S}\right\},$$

$$\phi(\boldsymbol{\mu}) = \frac{P}{K}\mathbf{1} + \boldsymbol{Q}\boldsymbol{\mu}.$$

As the preimage of an open set under a continuous function, $\mathcal{U}$ is an open subset of $\mathbb{R}^{K-1}$. Moreover, the map ϕ is a C^∞ diffeomorphism, see also Lemma 2.3.4. Accordingly, the map $\boldsymbol{r} \circ \phi$ is a differentiable global parameterization of $\mathcal{M}$. Global parameterizations of the Pareto set play an important role in the application of the solution method developed in Sect. 3.5. Section 3.5.3 provides a generic method for constructing a global parameterization. However, using a generic parameterization is not necessary in the MISO broadcast channel with linear precoding. In this case, the map $\boldsymbol{r} \circ \phi$ is a *natural parameterization*, which is very convenient to use due to the fact that $(\boldsymbol{r} \circ \phi)(\boldsymbol{\mu})$ can be computed in closed-form. The availability of a closed-form parameterization represents a special case. In general, it is not possible to find a parameterization of the Pareto set in closed-form. As an example, for the MIMO BC with dirty paper coding considered in Sect. 4.2, a closed-form parameterization is not available.

4.2 MIMO Broadcast Channel with Dirty Paper Coding

Again, a broadcast channel with K receivers is considered. The transmitter has N transmit antennas, while receiver k is equipped with M_k receive antennas. The transmitter sends independent information to each of the receivers.

The received signal at receiver k is given by

$$\boldsymbol{y}_k = \boldsymbol{H}_k \sum_{i=1}^{K} \boldsymbol{x}_i + \boldsymbol{\eta}_k,$$

where $\boldsymbol{H}_k \in \mathbb{C}^{M_k \times N}$ is the channel to receiver k and $\boldsymbol{x}_k \in \mathbb{C}^N$ is the signal transmitted to receiver k. Furthermore, $\boldsymbol{\eta}_k$ is the circularly symmetric complex

Gaussian noise at receiver k, with zero-mean and covariance matrix $\boldsymbol{I}$. The transmitted signal

$$\boldsymbol{x} = \sum_{k=1}^{K} \boldsymbol{x}_k$$

is subject to a sum-power constraint

$$\mathrm{E}\left[\|\boldsymbol{x}\|_2\right] \leq P. \tag{4.28}$$

4.2.1 PHY Layer Parameterization

Let $\boldsymbol{Q}_k$ denote the transmit covariance matrix of user k,

$$\boldsymbol{Q}_k = \mathrm{E}\left[\boldsymbol{x}_k \boldsymbol{x}_k^{\mathrm{H}}\right].$$

Under the assumption that $\boldsymbol{x}_k$ and $\boldsymbol{x}_i$ are statistically independent for $k \neq i$, the sum-power constraint (4.28) implies a trace constraint on the sum of the users' covariance matrices:

$$\mathrm{tr}\left(\sum_{k=1}^{K} \boldsymbol{Q}_k\right) \leq P.$$

Accordingly, with

$$\boldsymbol{Q} = (\boldsymbol{Q}_1, \dots, \boldsymbol{Q}_K),$$

the set of feasible transmit covariance matrices is given by

$$\mathcal{Q} = \left\{\boldsymbol{Q} : \boldsymbol{Q}_k \in \mathbb{H}_+^N, \mathrm{tr}\left(\sum_{k=1}^{K} \boldsymbol{Q}_k\right) \leq P\right\}.$$

where $\mathbb{H}_+^N$ denotes the set of positive semidefinite Hermitian $N \times N$ matrices.

As proved in [74], capacity is achieved by *dirty paper coding* (DPC). Let π denote the encoding order, i.e., $\pi : \{1, \dots, K\} \to \{1, \dots, K\}$ is a permutation and $\pi(i)$ is the index of the user which is encoded at the i-th position. Moreover, let Π denote the set of all possible permutations on $\{1, \dots, K\}$.

For fixed $\boldsymbol{Q}$ and π, an achievable rate vector is given by $\boldsymbol{r}(\boldsymbol{Q}, \pi) = (r_1(\boldsymbol{Q}, \pi), \dots, r_K(\boldsymbol{Q}, \pi))$, with

$$r_{\pi(i)}(\boldsymbol{Q}, \pi) = \log_2 \frac{\det\left(\boldsymbol{I} + \boldsymbol{H}_{\pi(i)}(\sum_{j \geq i} \boldsymbol{Q}_{\pi(j)})\boldsymbol{H}_{\pi(i)}^{\mathrm{H}}\right)}{\det\left(\boldsymbol{I} + \boldsymbol{H}_{\pi(i)}(\sum_{j > i} \boldsymbol{Q}_{\pi(j)})\boldsymbol{H}_{\pi(i)}^{\mathrm{H}}\right)}. \tag{4.29}$$

In the general model defined in Chap. 2, the physical layer is specified by a set of feasible parameter setups $\mathcal{X}_1$ and a rate map $\boldsymbol{r}$ that assigns a rate vector $\boldsymbol{r}(\boldsymbol{x}_1) \in \mathbb{R}_+^K$ to each $\boldsymbol{x}_1 \in \mathcal{X}_1$. In case of a MIMO broadcast channel, these entities are specified by

$$\mathcal{X}_1 = \{\boldsymbol{x}_1 = (\boldsymbol{Q}, \pi), \boldsymbol{Q} \in \mathcal{Q}, \pi \in \Pi\} = \mathcal{Q} \times \Pi,$$

$$\boldsymbol{r} : \mathcal{Q} \times \Pi \rightarrow \mathbb{R}_+^K, (\boldsymbol{Q}, \pi) \mapsto \boldsymbol{r}(\boldsymbol{Q}, \pi),$$

with $\boldsymbol{r}(\boldsymbol{Q}, \pi)$ as defined in (4.29).

Under the assumption that $P > 0$, the set $\mathcal{X}_1$ is compact and nonempty, and the rate function $\boldsymbol{r}$ is continuous. Accordingly, the pair $(\boldsymbol{r}, \mathcal{X}_1)$ fulfills the requirements of Definitions 2.1.1 and 2.2.1.

4.2.2 Rate Region

Given a physical layer specification $(\boldsymbol{r}, \mathcal{X}_1)$, the rate region is defined by

$$\mathcal{R} = \boldsymbol{r}(\mathcal{X}_1),$$

cf. Sect. 2.3. As discussed in Sect. 2.3.5, by adding a scheduling mode, each point in the convex hull of $\mathcal{R}$ can be achieved by time sharing. In the context of MIMO broadcast channels, $\mathcal{C} = \operatorname{conv}(\mathcal{R})$ is prominently known as the *capacity region* of the MIMO BC [74].

Throughout this monograph, it is assumed that rate regions are proper sets. For a MIMO BC with DPC, this fundamental property is established by Propositions 4.2.1 and 4.2.2.

Proposition 4.2.1. *The rate region $\mathcal{R} = \boldsymbol{r}(\mathcal{Q} \times \Pi)$ is comprehensive.*

Proof. Let $\boldsymbol{s}' \in \mathcal{R}$. Accordingly, there exists a pair $(\boldsymbol{Q}', \pi')$ such that $\boldsymbol{s}' = \boldsymbol{r}(\boldsymbol{Q}', \pi')$. Define a function $\boldsymbol{F} : \mathbb{R}_+^K \rightarrow \times_{k=1}^K \mathbb{H}_+^N$ as follows:

$$\boldsymbol{F}(\boldsymbol{p}) = (\boldsymbol{F}_1(\boldsymbol{p}), \ldots, \boldsymbol{F}_K(\boldsymbol{p})),$$

with

$$\boldsymbol{F}_k(\boldsymbol{p}) = p_k \frac{\boldsymbol{Q}'_k}{\operatorname{tr}\left(\boldsymbol{Q}'_k\right)}.$$

Note that $\operatorname{tr}\left(\sum_{k=1}^{K} \boldsymbol{F}_k(\boldsymbol{p})\right) = \mathbf{1}^{\mathrm{T}}\boldsymbol{p}$. By using similar arguments as in Proposition 4.1.1, it can be shown that given $\boldsymbol{s}$, with $\mathbf{0} \leq \boldsymbol{s} \leq \boldsymbol{s}'$, there exists $\boldsymbol{p} \in \mathbb{R}_+^K$ with $\mathbf{1}^{\mathrm{T}}\boldsymbol{p} \leq P$ such that

$$\boldsymbol{r}(\boldsymbol{F}(\boldsymbol{p}), \pi') = \boldsymbol{s},$$

which implies that $\boldsymbol{s} \in \mathcal{R}$. □

Proposition 4.2.2. *The rate region* $\mathcal{R} = \boldsymbol{r}(\mathcal{Q} \times \Pi)$ *is proper if and only if* $P > 0$ *and*

$$\boldsymbol{H}_k \neq \mathbf{0}, \forall k.$$

Proof. $\mathcal{R}$ is the image of a compact set under a continuous function, thus $\mathcal{R}$ is compact. By Proposition 4.2.1, $\mathcal{R}$ is comprehensive. The converse follows by letting

$$\boldsymbol{Q}_k = \frac{P}{K} \frac{\boldsymbol{H}_k^{\mathrm{H}} \boldsymbol{H}_k}{\operatorname{tr}\left(\boldsymbol{H}_k^{\mathrm{H}} \boldsymbol{H}_k\right)}, k \in \{1, \ldots, K\}.$$

On the other hand, if either $\boldsymbol{H}_k = \mathbf{0}$ or $P = 0$, then $r_k(\boldsymbol{Q}, \pi) = 0, \forall (\boldsymbol{Q}, \pi) \in \mathcal{Q} \times \Pi$. □

4.2.3 Weighted Sum Rate Maximization

The rate map $\boldsymbol{r}$ defined in (4.29) is a nonconcave function of the covariance matrices $\boldsymbol{Q}$, and the set of encoding orders is discrete. As a result, a utility maximization problem in the MIMO BC in parameter space formulation generally represents a nonconvex optimization problem, even if time sharing between different parameter setups is possible (the corresponding rate map $\boldsymbol{r}'$ is still nonconcave in $\boldsymbol{Q}' = (\boldsymbol{Q}^1, \ldots, \boldsymbol{Q}^K)$, and the optimization variable $\pi' = (\pi^1, \ldots, \pi^K)$ is still discrete). Fortunately, there exists a problem that can be solved efficiently in the MIMO BC, namely *weighted sum rate maximization*, with $u(\boldsymbol{s}) = \boldsymbol{\lambda}^{\mathrm{T}}\boldsymbol{s}$ and $\boldsymbol{\lambda} \in \mathbb{R}_+^K$. As discussed before, in case of weighted sum rate maximization, time sharing between parameter setups is not needed to achieve the maximum. Moreover, for this particular utility function, an optimal encoding order π^* can be inferred from $\boldsymbol{\lambda}$ [75–77]. As a result, the problem is reduced to finding the optimal transmit covariance matrices, which can be solved as a convex problem in the dual MAC [77]. Based on this result, a number of efficient algorithms for solving the weighted sum rate maximization problem in the MIMO BC have been proposed [78–80]. The convex structure of the weighted sum rate maximization problem is a result of the ability to cancel known interference by dirty paper coding [75, 76].

The ability to efficiently solve weighted sum rate maximization problems is the key to applying the solution methods from Chap. 3. As discussed in Sect. 3.3, a concave utility function u can be maximized over a convex rate region $\mathcal{C}$ by applying Lagrange duality, provided a means for solving weighted sum rate maximization problem over $\mathcal{C}$ is available. In addition, a nonconcave utility function can be maximized by using the polyblock algorithm discussed in Sect. 3.4. The polyblock algorithm relies on the ability to solve intersection problems. As shown by Proposition 4.2.1, both $\mathcal{R}$ and $\mathcal{C}$ are proper, hence an intersection problem on $\mathcal{C}$ is equivalent to a generalized max–min problem on $\mathcal{C}$, cf. Sect. 5.1.3. The generalized max–min problem on $\mathcal{C}$ is again a convex rate space problem, which can be solved by means of Lagrange duality and weighted sum rate maximization, as discussed in Sect. 3.3. Finally, the local method from Sect. 3.5 needs a method for solving an intersection problem and a method for computing a tangent space basis at a given point on the Pareto boundary of the rate region. As shown in Sect. 4.2.4, a tangent space basis at a point on the Pareto boundary of $\mathcal{C}$ can also be obtained from a weighted sum rate maximization problem. To summarize, the availability of efficient algorithms for solving weighted sum rate maximization problems enables application of all three solution methods for utility maximization over $\mathcal{C}$. Importantly, only maximization over the convex capacity region $\mathcal{C}$ is possible. The rate region $\mathcal{R}$ is generally nonconvex. As all solution methods that build upon weighted sum rate maximization are based on Lagrange duality, a convex rate region is required, as otherwise strong duality need not hold.

Without dirty paper coding, the weighted sum rate maximization problem is nonconvex, even in the dual MAC. Hence, for multi-antenna broadcast channels with linear precoding only (i.e., *without* DPC), no efficient[6] solutions for the weighted sum rate maximization problem exist, cf. Sects. 4.1 and 6.1. In other words, dirty paper coding is not only necessary to achieve capacity, it also allows for efficient solutions to the class of concave utility maximization problems.

4.2.4 Global Parameterization of the Pareto Boundary

In Sect. 3.5, a method for computing a stationary point on the Pareto boundary of the rate region is developed. By using a global parameterization, the problem of finding a stationary point can be transformed into a standard constrained optimization problem. As discussed in Sect. 4.1.5, for the MISO broadcast channel with linear precoding, a global parameterization that covers the whole Pareto set is available in closed-form. Unfortunately, such a closed-form parameterization is not available for the Pareto set of the capacity region $\mathcal{C}$ of the MIMO broadcast channel. Instead, the generic parameterization developed in Sect. 3.5.3 can be used. The

---

[6]Here, the term "efficient solution" refers to solution whose computational complexity is practically feasible for large K.

generic parameterization does not cover the whole Pareto set. Instead, it is assumed that par$(\mathcal{C})$ can be decomposed into a union of disjoint differentiable manifolds $\mathcal{M}_{\mathcal{D}}$, with each manifold having a global parameterization, cf. Sect. 3.5.5. In the following, it is assumed that a stationary point on $\mathcal{M} = \mathcal{M}_{\emptyset}$ is computed. As shown in Sect. 3.5.5, operating on $\mathcal{M}_{\mathcal{D}}$, with $\mathcal{D} \supset \emptyset$, is equivalent to operating on $\mathcal{M}_{\emptyset}$, only the dimension of the problem is reduced.

In order to evaluate a parameterization, two methods are needed: a method for solving an intersection problem and a method for computing a tangent space basis at a given point on the Pareto boundary of the capacity region. As discussed in Sect. 4.2.3, an intersection problem on $\mathcal{C}$ can be solved based on Lagrange duality and a method for solving weighted sum rate maximization problems. Hence, the only open issue that remains is the computation of a tangent space basis at a given point $\boldsymbol{s} \in \mathcal{M}$.

Assume that a vector $\boldsymbol{\lambda} \in \mathbb{R}_{+}^{K} \setminus \{\mathbf{0}\}$ is known such that $\boldsymbol{s} \in \mathcal{M}$ is a maximizer of a weighted sum rate maximization problem over $\mathcal{C}$ with weight $\boldsymbol{\lambda}$, i.e.,

$$\boldsymbol{\lambda}^{\mathrm{T}}\boldsymbol{s} = \max_{\boldsymbol{z}\in\mathcal{C}} \boldsymbol{\lambda}^{\mathrm{T}}\boldsymbol{z}.$$

By Proposition A.5.6, such a vector $\boldsymbol{\lambda}$ always exists. From $\boldsymbol{s} \in \mathcal{M}$ and $\mathcal{M} \subset \mathcal{C}$ it follows that

$$\boldsymbol{\lambda}^{\mathrm{T}}\boldsymbol{s} = \max_{\boldsymbol{z}\in\mathcal{M}} \boldsymbol{\lambda}^{\mathrm{T}}\boldsymbol{z},$$

which implies that $\boldsymbol{s}$ is a stationary point of the weighted sum rate maximization problem in $\mathcal{M}$. Hence,

$$\boldsymbol{\lambda}^{\mathrm{T}}\boldsymbol{v} \leq 0, \forall \boldsymbol{v} \in \mathcal{T}_{\mathcal{M}}(\boldsymbol{s}).$$

By Proposition A.8.3, $\mathcal{T}_{\mathcal{M}}(\boldsymbol{s})$ is a $K-1$-dimensional subspace, hence

$$\mathcal{T}_{\mathcal{M}}(\boldsymbol{s}) = \mathrm{null}\left(\boldsymbol{\lambda}^{\mathrm{T}}\right).$$

As a result, if the value of $\boldsymbol{\lambda}$ is known, a tangent space basis at $\boldsymbol{s}$ can be obtained by computing a basis of the nullspace of $\boldsymbol{\lambda}^{\mathrm{T}}$.

In the iterative algorithm developed in Sect. 3.5, the point $\boldsymbol{s}$ is obtained by evaluating the parameterization at the previous iterate $\boldsymbol{\mu}$. Hence, $\boldsymbol{s}$ is computed by solving an intersection problem on $\mathcal{C}$ using Lagrange duality. Accordingly, the value of $\boldsymbol{\lambda}$ corresponding to $\boldsymbol{s}$ is just the optimal value of the dual variable in the corresponding dual problem (3.22). In other words, the tangent space basis is a byproduct of solving the intersection problem via Lagrange duality.

Evidently, in case of a MIMO BC with DPC, evaluating the global parameterization of $\mathcal{C}$ is considerably more complex than evaluating the global parameterization of $\mathcal{R}$ in the case of a MISO BC with linear precoding. Still, all subproblems (intersection problem and weighted sum rate maximization problem) are convex optimization problems. As a result, computing a stationary point in $\mathcal{C}$ is practically feasible for large K, while computing a globally optimal rate vector by means of monotonic optimization is practically feasible for relatively small K only. Hence, in the context of a MIMO BC with DPC, local methods are attractive in case of a nonconcave utility function. On the other hand, if the utility function is concave, a single iteration of the local method is of similar complexity as computing the optimal solution by means of Lagrange duality, as each iteration of the local method involves solving a generalized max–min problem, which is itself a concave utility maximization problem on $\mathcal{C}$.

Chapter 5
Utility Models

In Chap. 2, a generic utility maximization problem is formulated. By Definition 2.2.1, the system utility function u can be any continuous increasing function. Chapter 3 discusses various methods for solving the utility maximization problem. Two of the proposed methods make further assumptions concerning the properties of the function u: for the dual method, concavity of u is assumed, whereas the local method assumes continuous differentiability of u. As long as the corresponding basic assumptions are met, the solution methods are applicable to any utility function u.

Utility maximization is a popular concept for formulating resource allocation problems in wired and wireless networks, cf. [3, 5, 13, 14, 18, 56, 81, 82] and references therein. In order to illustrate possible choices for the system utility function u, this chapter collects some of the most frequently used utility models from the literature.

A common model for network traffic is that of flows (or equivalently, end-to-end connections) with infinite backlog. Utility is then typically a function of the end-to-end flow rates. In contrast, throughout this work, utility is a function of the link rate vector $\boldsymbol{s}$. Under the assumption that there exists a function $\boldsymbol{f}$ that maps a link rate vector $\boldsymbol{s}$ into a vector of end-to-end flow rates $\boldsymbol{y}$, the utility function u represents a generalization of the flow-level utility concept. Let B denote the number of flows. The end-to-rate flow rate of flow b is given by $y_b = f_b(\boldsymbol{s})$. Typically, a utility value $u_b(y_b)$ is assigned to each flow rate y_b by using a utility function u_b. The resulting utility vector $\boldsymbol{v} = \boldsymbol{u}(\boldsymbol{y})$ is then mapped into an overall scalar utility value by an increasing function σ. Common examples for the function σ are

$$\sigma(\boldsymbol{v}) = \mathbf{1}^{\mathrm{T}}\boldsymbol{v},$$

and

$$\sigma(\boldsymbol{v}) = \min_{b\in\{1,\ldots,B\}} v_b.$$

In both cases, σ is continuous, increasing, and concave.

J. Brehmer, *Utility Maximization in Nonconvex Wireless Systems*, Foundations in Signal Processing, Communications and Networking 5, DOI 10.1007/978-3-642-17438-4_5,

To summarize, the following model for the system utility u results:

$$u = \sigma \circ \boldsymbol{u} \circ \boldsymbol{f},$$
$$u(\boldsymbol{s}) = \sigma(\boldsymbol{u}(\boldsymbol{f}(\boldsymbol{s}))).$$

Under the assumption that the functions $\boldsymbol{f}, \boldsymbol{u}$, and σ are continuous and increasing, the function u is a valid utility function according to Definition 2.2.1. Furthermore, if, in addition to being increasing, $\boldsymbol{f}, \boldsymbol{u}$, and σ are concave, then u is concave.

As mentioned above, the b-th component function u_b of $\boldsymbol{u}$ only depends on the flow rate of flow b, i.e., $v_b = u_b(y_b)$. In the following, possible choices for the utility function u_b are discussed. Moreover, for simplicity, it is assumed that the map from link rates to flow rates is the identity map, i.e., $B = K$ and

$$f_k(s_k) = s_k, k = 1, \dots, K.$$

5.1 Concave Utility Functions

In most works on utility-based resource allocation, a basic assumption is the concavity of the utility functions u_k [14, 38]. Accordingly, there exists a wide array of examples for concave utility functions. Some of the most popular models are briefly reviewed in this section.

5.1.1 Sum-Throughput and Max–Min Fairness

Combining the simple choice

$$u_k(s_k) = s_k, k = 1, \dots, K \tag{5.1}$$

with the two choices for σ yields two fundamental (and antipodal) objectives for resource allocation: sum-throughput and max–min fairness [81].

5.1.2 Logarithmic Utility Functions

Evidently, in (5.1) a linear increase in s_k leads to a linear increase in $u_k(s_k)$, regardless of the value of s_k. This behaviour is not consistent with the law of diminishing returns. A family of utility functions u_k that yields diminishing marginal returns is the class of increasing and strictly concave functions. With respect to resource allocation problems, the most popular elements in this family are the logarithmic utility functions. Common examples are

$$u_k(s_k) = \ln(s_k), \tag{5.2}$$

and

$$u_k(s_k) = \ln(1 + s_k). \tag{5.3}$$

Both logarithmic utility functions are usually used in conjunction with $\sigma(\boldsymbol{v}) = \mathbf{1}^{\mathrm{T}}\boldsymbol{v}$. Despite their similarity, they can lead to very different resource allocations. This behavior stems from the fact that using the utility function (5.2) for link k enforces a non-zero rate on this link, while the function (5.3) allows for zero-rate allocations. The flexibility to allow for zero rates becomes particularly important in overloaded wireless systems, where the number of degrees of freedom available for separating links is smaller than the actual number of links, which causes strong interference if all links are active. Strong interference usually translates into a pronounced nonconvexity of the rate region. An example for such an overloaded system is given in Sect. 6.1.1.

If the rate region is convex, the rate vector that is optimal for a utility model $\sigma(\boldsymbol{v}) = \mathbf{1}^{\mathrm{T}}\boldsymbol{v}$ and $u_k(s_k) = \ln(s_k), k = 1, \ldots, K$, is *proportionally fair* [5]. It is important to note that proportional fairness is a property of a rate vector. For convex rate regions, a proportional fair rate vector can be obtained by solving a utility maximization problem. For nonconvex rate regions, however, there may not even exist a rate vector that is proportionally fair according to Kelly's definition [46]. In contrast, a utility maximization problem that fulfills the requirements of Definition 2.2.1 always has a maximizer, cf. Proposition 2.1.1. In other words, using the utility function $u_k(s_k) = \ln(s_k)$ is possible regardless of the properties of the rate region, while the maximizer of the corresponding rate space problem is not necessarily proportional fair.

Another set of concave utility functions are the so-called α-fair utilities [56]. For a convex rate region, their elements provide different trade-offs between max–min fairness and proportional fairness, depending on the choice of a parameter α. The application of α-fair utilities to nonconvex systems is discussed in [83].

5.1.3 Generalized Max–Min

The max–min fair solution can be generalized by letting

$$u_k(s_k) = \frac{s_k - s'_k}{d_k}, k = 1, \ldots, K,$$

with $\boldsymbol{s}' \in \mathbb{R}^K_+$ and $\boldsymbol{d} \in \mathbb{R}^K_{++}$. Given a rate region $\mathcal{R}$, the *generalized max–min problem* is then defined as

$$\max_{s \in \mathcal{R}} \min_k \frac{s_k - s'_k}{d_k}. \tag{5.4}$$

Clearly, the generalized max–min problem corresponds to a utility maximization problem with utility function

$$u(s) = \min_k \frac{s_k - s'_k}{d_k}.$$

As the point-wise minimum over a family of affine functions, u is concave.

In Sect. 2.3.2, the so-called intersection problem is defined as follows:

$$\max_{\beta \in \mathbb{R}} \beta \quad \text{s.t.} \quad (s' + \beta d)^+ \in \mathcal{R}.$$

Proposition 5.1.1 shows that for proper rate regions, the generalized max–min problem and the intersection problem are equivalent.

Proposition 5.1.1. *Let $\mathcal{R}$ denote a proper subset of $\mathbb{R}_+^K$. Then the generalized max–min problem and the intersection problem have the same optimum, i.e.,*

$$\max_{s \in \mathcal{R}} \min_k \frac{s_k - s'_k}{d_k} = \max_\beta \beta \quad \textit{s.t.} \quad (s' + \beta d)^+ \in \mathcal{R}.$$

Proof. A proper set is nonempty, hence the maxima of both problems are well-defined. Let β_{GM} denote the maximum of the generalized max–min problem, while β_{IP} is the maximum of the intersection problem. By definition of β_{GM}, there exists $s \in \mathcal{R}$ such that

$$\beta_{\text{GM}} = \min_k \frac{s_k - s'_k}{d_k}.$$

Re-arranging yields

$$s_k \geq s'_k + \beta_{\text{GM}} d_k, \forall k.$$

Taking the $(\cdot)^+$ operator on both sides and observing that $(s)^+ = s$ for any $s \in \mathcal{R}$ yields

$$s \geq (s' + \beta_{\text{GM}} d)^+ \geq \mathbf{0}.$$

By assumption, $\mathcal{R}$ is comprehensive, hence $(s' + \beta_{\text{GM}} d)^+ \in \mathcal{R}$, which implies $\beta_{\text{GM}} \leq \beta_{\text{IP}}$.

By definition of β_{IP}, $(s' + \beta_{\text{IP}} d)^+ \in \mathcal{R}$, hence there exists $s \in \mathcal{R}$ such that

$$s \geq s' + \beta_{\text{IP}} d.$$

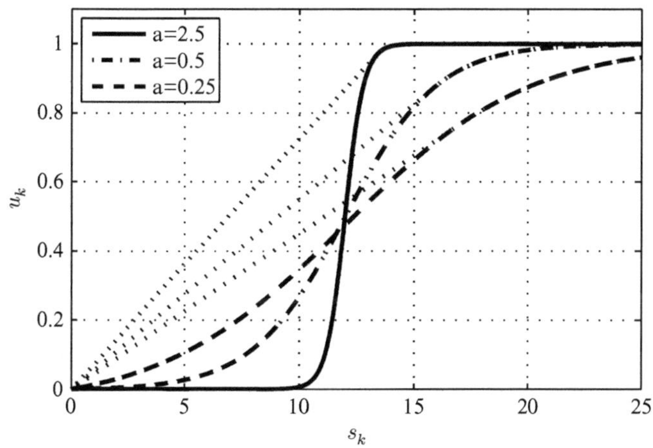

Fig. 5.1 Sigmoidal utility function, $b = 12$

Re-arranging yields

$$\frac{s_k \quad s'_k}{d_k} \geq \beta_{\mathrm{IP}},$$

which implies $\beta_{\mathrm{GM}} \geq \beta_{\mathrm{IP}}$. □

5.2 Nonconcave Utility Functions

Nonconcave utility functions have been introduced in the context of *application-oriented utilities* [6, 7, 57]. They are motivated by the observation that real-time and multimedia application can be modeled more precisely by nonconcave utility functions. One frequently used class of nonconcave utility functions are the sigmoidal functions. Sigmoidal functions are strictly increasing, strictly convex for rates smaller below the inflection point, and strictly concave above the inflection point. In [39, 44, 57], the following sigmoidal function is used:

$$u_k(s_k) = c\left(\frac{1}{1 + \exp(-a(s_k - b))} + d\right). \tag{5.5}$$

The steepness of the transition between the minimum and the maximum utility value is controlled by the parameter a, whereas b determines the inflection point of the utility curve. The constants c and d are used to normalize u_k such that $u_k(0) = 0$ and $u_k(\infty) = 1$. Figure 5.1 shows $u_k(s_k)$ for $b = 12$ and different values of a. Also shown are the corresponding concave envelopes (dotted lines).

Chapter 6
Case Studies

Based on the general definition of a utility maximization problem from Chaps. 2, 4 and 5 introduce models for the physical layer and the utility layer. By choosing a specific model for both physical layer and utility layer, a realization of a utility maximization problem results, represented by the data vector $(u, h, \boldsymbol{r}, \mathcal{X}_1)$. Chapter 3 discusses methods to solve a given utility maximization problem. The applicability of each of the discussed methods depends on the properties of the utility maximization problem at hand. In this chapter, the application of the general framework developed in Chaps. 2 and 3 is demonstrated. By combining the models from Chaps. 4 and 5, a set of utility maximization problems is obtained. Each problem is solved using the methods provided in Chap. 3. It should be noted, however, that the case studies presented in this chapter represent only a subset of the problems to which the proposed framework can be applied, due to the generality of the developed methods.

6.1 Utility Maximization in the MISO Broadcast Channel with Linear Precoding

In the first case study, the physical layer model is a MISO broadcast channel with linear precoding. This physical layer model is described in detail in Sect. 4.1. Two physical layer configurations are considered: In the first configuration, the base station manages interference by beamforming (spatial signal processing). Accordingly, the physical layer is specified by the pair $(\boldsymbol{r}, \mathcal{X}_1)$ defined by (4.4) and (4.5). In the second configuration, time sharing between parameter setups is possible, i.e., the base station performs joint beamforming and scheduling. The physical layer is specified by a pair $(\boldsymbol{r}', \mathcal{X}_1')$, where $(\boldsymbol{r}', \mathcal{X}_1')$ is derived from $(\boldsymbol{r}, \mathcal{X}_1)$ as described in Sect. 2.3.5.

J. Brehmer, *Utility Maximization in Nonconvex Wireless Systems*, Foundations in Signal Processing, Communications and Networking 5, DOI 10.1007/978-3-642-17438-4_6,

Under mild assumptions, the rate region of the MISO BC with linear precoding is proper, as shown by Proposition 4.1.2. Without time sharing, however, the rate region is not necessarily convex.

In the following, let $\mathcal{R}$ denote the rate region without scheduling and $\mathcal{C}$ the rate region with scheduling, i.e., $\mathcal{R}$ is the rate region associated with $(\boldsymbol{r}, \mathcal{X}_1)$, while $\mathcal{C}$ is the rate region associated with $(\boldsymbol{r}', \mathcal{X}_1')$.

Two different utility models (u, h) are considered. In both models, $h(\mathbf{0}) = 0$, i.e., there is no additional constraint at the utility level. For the first model, it is assumed that u is concave, while for the second model, u is nonconcave.

Combining two physical layer configurations with two utility models yields four utility maximization problems:

(1) Concave u, no scheduling.
(2) Nonconcave u, no scheduling.
(3) Concave u, with scheduling.
(4) Nonconcave u, with scheduling.

The parameter space formulation of each of the four problems results in a nonconvex optimization problem, as discussed in Sect. 6.1.1. In Sect. 6.1.2, it is shown how the globally optimal parameter setup can be determined, using the methods from Chap. 3.

6.1.1 Parameter Space Formulation

Given a continuous and increasing utility function u, the parameter space problem for the physical layer $(\boldsymbol{r}, \mathcal{X}_1)$ is given by

$$\max_{\boldsymbol{w} \in \mathbb{C}^{KN}} u(\boldsymbol{r}(\boldsymbol{w})) \quad \text{s.t.} \quad \|\boldsymbol{w}\|_2^2 \leq P. \tag{6.1}$$

Likewise, the parameter space problem with scheduling reads as

$$\begin{aligned} \max_{(\boldsymbol{b}, \boldsymbol{w}^1, \ldots, \boldsymbol{w}^K)} u\left(\sum_{k=1}^{K} b_k \boldsymbol{r}(\boldsymbol{w}^k)\right) \quad \text{s.t.} \quad & \boldsymbol{w}^k \in \mathbb{C}^{KN}, k = 1, \ldots, K, \\ & \|\boldsymbol{w}^k\|_2^2 \leq P, k = 1, \ldots, K, \\ & \boldsymbol{b} \in \mathbb{R}_+^K, \mathbf{1}^{\mathrm{T}} \boldsymbol{b} = 1. \end{aligned} \tag{6.2}$$

In general, neither problem (6.1) nor problem (6.2) are convex problems, due to the nonconcavity of the rate map $\boldsymbol{r}$.

Moreover, the parameter space formulation does not preserve the monotonicity structure that is present at the utility level. While u is increasing, neither $\boldsymbol{r}$ nor $\boldsymbol{r}'$ is increasing. As a result, both problem (6.1) and problem (6.2) generally represent nonconvex optimization problems with no structure except for the basic properties implied by Definition 2.2.1.

6.1.2 Globally Optimal Solutions

The key to obtaining globally optimal solutions to the utility maximization problems $(u, \boldsymbol{r}, \mathcal{X}_1)$ and $(u, \boldsymbol{r}', \mathcal{X}_1')$ is the rate space formulation, cf. Sect. 3.2. If no scheduling mode is provisioned, the rate space problem for $(u, \boldsymbol{r}, \mathcal{X}_1)$ is given by

$$\max_{\boldsymbol{s}} u(\boldsymbol{s}) \quad \text{s.t.} \quad \boldsymbol{s} \in \mathcal{R}. \tag{6.3}$$

Problem (6.3) corresponds to the maximization of an increasing function over a compact comprehensive set. Even if u is assumed to be concave, problem (6.3) can in general not be assumed to be convex, due to the potential nonconvexity of $\mathcal{R}$, cf. Sect. 4.1.2.

With scheduling, the rate space problem reads as

$$\max_{\boldsymbol{s}} u(\boldsymbol{s}) \quad \text{s.t.} \quad \boldsymbol{s} \in \mathcal{C}. \tag{6.4}$$

Problem (6.4) corresponds to the maximization of an increasing function over a convex comprehensive set. If u is concave, problem (6.4) is convex.

Accordingly, the four utility maximization problems can be categorized into three problem classes:

- Type 1: maximization of an increasing function over a comprehensive set $\mathcal{R}$ (monotonic optimization problem).
- Type 2: maximization of a concave function over a convex set $\mathcal{C}$ (convex optimization problem).
- Type 3: maximization of a nonconcave, increasing function over a convex set $\mathcal{C}$ (monotonic optimization problem).

While types 1 and 3 seem similar, there is a major difference: While the physical layer provides an efficient algorithm to solve an intersection problem over $\mathcal{R}$, cf. Sect. 4.1.3, there is no algorithm available for directly solving an intersection problem over $\mathcal{C}$. This property leads to a significantly more complex solution for problems of type 3, as shown in the following.

Regardless whether u is concave or not, Problem (6.3) is a problem of type 1. It is solved within the framework of monotonic optimization, using the methods developed in Sect. 3.4. There is no additional increasing constraint, and $\mathcal{R} \not\subset \mathbb{R}^K_{++}$, thus the extended polyblock algorithm from Sect. 3.4.3 is employed to compute an optimal rate vector. In each iteration of the polyblock algorithm, a vertex of the polyblock has to be projected on the weak Pareto boundary of $\mathcal{R}$. The corresponding intersection problem is solved using the algorithm developed in

Sect. 4.1.3. As a result, by combining the extended polyblock algorithm from Sect. 3.4.3 with the algorithm for solving the intersection problem from Sect. 4.1.3, the rate space problem (6.3) can be solved to global optimality for any continuous increasing function u.

A special case of problem (6.3) is weighted sum rate maximization, with

$$u(\boldsymbol{s}) = \boldsymbol{\lambda}^{\mathrm{T}} \boldsymbol{s},$$

and $\boldsymbol{\lambda} \in \mathbb{R}_{+}^{K}$. Moreover, according to Corollary A.5.9,

$$\max_{\boldsymbol{s} \in \mathcal{R}} \boldsymbol{\lambda}^{\mathrm{T}} \boldsymbol{s} = \max_{\boldsymbol{s} \in \mathcal{C}} \boldsymbol{\lambda}^{\mathrm{T}} \boldsymbol{s}.$$

As a consequence, weighted sum rate maximization over $\mathcal{R}$ and $\mathcal{C}$ can be solved to global optimality with the methods developed in Sect. 3.4. Weighted sum rate maximization plays a key role in solving problems of type 2 and 3.

If u is concave, problem (6.4) is a problem of type 2. Being a convex problem, it is solved using the Lagrange duality-based solution approach developed in Sect. 3.3. The duality-based approach requires a method to determine a maximizer of a weighted sum of rates over $\mathcal{C}$. Such a maximizer can be obtained by solving a problem of type 1, using the methods from Sects. 3.4 and 4.1.3. At this point, it becomes evident that introducing a scheduling mode complicates the solution. Without scheduling, a single problem of type 1 has to be solved. With scheduling, a problem of type 1 has to be solved for each evaluation of the dual function. The dual function is minimized by an iterative algorithm, which generally requires a significant number of function evaluations.

If u is nonconcave, problem (6.4) is a problem of type 3. It is a monotonic optimization problem, and is solved using the methods from Sect. 3.4. Analogous to the solution of problem (6.3), the extended polyblock algorithm requires a method to solve an intersection problem, but in this case, the intersection point with the weak Pareto boundary of $\mathcal{C}$ instead of $\mathcal{R}$ has to be computed. Solving the intersection problem over $\mathcal{C}$ is significantly harder than solving the intersection problem over $\mathcal{R}$. As pointed out before, while the physical layer provides an efficient algorithm to solve the intersection problem over $\mathcal{R}$, cf. Sect. 4.1.3, there is no algorithm available for directly solving the intersection problem over $\mathcal{C}$. What is available, however, is a method to solve weighted sum rate maximization problems, as weighted sum rate maximization is a problem of type 1. Following the remarks from Sect. 3.4.4, for comprehensive rate regions the intersection problem is equivalent to a generalized max–min problem. Moreover, the generalized max–min problem over $\mathcal{C}$ is a problem of type 2. Accordingly, it can be solved by Lagrange duality, using the methods from Sect. 3.3. In other words, by choosing the Lagrange duality-based approach for solving the intersection problem, the intersection point is constructed from a set of maximizers of weighted sum rate maximization problems by means of primal reconstruction, cf. Sect. 3.3. Evidently, the solution procedure for problems of type 3 is the most complex, as it includes multiple problems of types 1 and 2: At each iteration of the polyblock algorithm, a problem of type 2 is generated. Each type 2 problem is solved by Lagrange duality, which in turn results in a set of type 1 problems.

The hierarchical structure of the three solution approaches is summarized in Fig. 6.1.

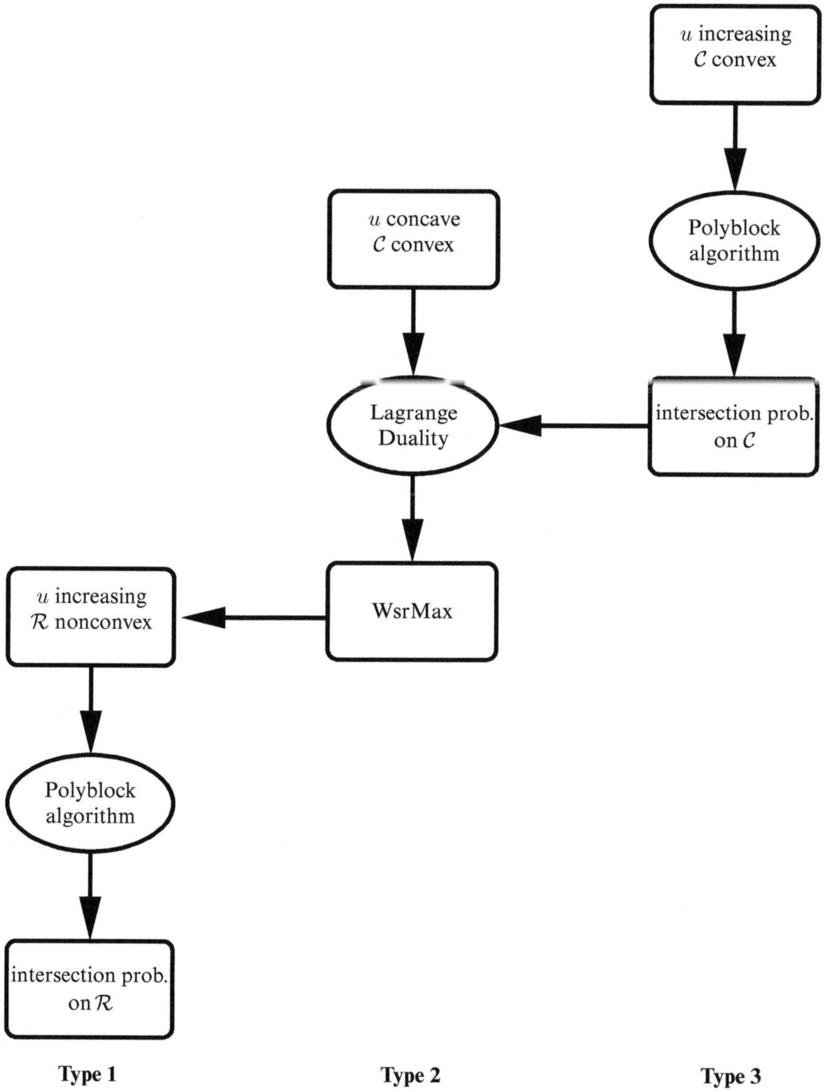

Fig. 6.1 Hierarchy of globally optimal solution methods

6.1.3 Local Solutions

A method for computing a local solution of the rate space problem $(u, \boldsymbol{h}, \mathcal{R})$ is proposed in Sect. 3.5. The local method computes a stationary point on the Pareto boundary of $\mathcal{R}$. As discussed in Sect. 3.5, it relies on the availability of a differentiable global parameterization of the Pareto boundary of the rate region $\mathcal{R}$. In Sect. 4.1.5, it is shown that for the Pareto set of the rate region of a MISO broadcast channel with linear precoding, a differentiable parameterization of the whole Pareto set is available. Moreover, this parameterization is given in closed-form, which makes the application of parameterization-based local methods particularly efficient. The parameterization $\boldsymbol{\psi} = \boldsymbol{r} \circ \boldsymbol{\phi}$ from Sect. 4.1.5 covers a superset of the Pareto boundary. In order to have $\boldsymbol{\psi}(\boldsymbol{\mu}) \in \operatorname{par}(\mathcal{R})$, the constraint $\boldsymbol{\psi}(\boldsymbol{\mu}) \geq \mathbf{0}$ is added. This corresponds to adding a constraint function $\boldsymbol{h} : \mathbb{R}^K \to \mathbb{R}^K$ with

$$\boldsymbol{h}(\boldsymbol{s}) = \boldsymbol{s}.$$

Using the parameterization $\boldsymbol{\psi}$, the utility maximization problem reads as

$$\max_{\boldsymbol{\mu} \in \mathbb{R}^{K-1}} u(\boldsymbol{\psi}(\boldsymbol{\mu})) \quad \text{s.t.} \quad \boldsymbol{\psi}(\boldsymbol{\mu}) \geq \mathbf{0}.$$

A KKT point of this problem can be obtained using standard methods, cf. Sect. 3.5.2. A more efficient formulation is obtained by recognizing that $\boldsymbol{\psi}(\boldsymbol{\mu}) \geq \mathbf{0}$ if and only if $\boldsymbol{\phi}(\boldsymbol{\mu}) \geq \mathbf{0}$, see Sect. 4.1.5. Accordingly, the nonlinear inequality constraint $\boldsymbol{\psi}(\boldsymbol{\mu}) \geq \mathbf{0}$ can be replaced by the linear inequality constraint

$$\frac{P}{K}\mathbf{1} + \boldsymbol{Q}\boldsymbol{\mu} \geq \mathbf{0}.$$

The results from Sect. 4.1.5 only apply to a MISO broadcast channel *without* scheduling. A closed-form parameterization of $\operatorname{par}(\mathcal{C})$ is not available. The generic parameterization proposed in Sect. 3.5.3 relies on two elementary methods: a method for solving an intersection problem and a method for computing a tangent space basis. As shown in Sect. 6.1.2, an intersection problem on $\mathcal{C}$ corresponds to a utility maximization problem of type 2, whose solution involves solving a sequence of monotonic optimization problems, cf. Fig. 6.1. In other words, obtaining a local solution on $\mathcal{C}$ is of similar complexity as obtaining a global solution. As a consequence, it does not make sense to use local methods that depend on a parameterization of $\operatorname{par}(\mathcal{C})$.

The key to good solutions with low complexity is to recognize that both solving an intersection problem on $\mathcal{C}$ and maximizing a concave utility over $\mathcal{C}$ require the solution of weighted sum rate maximization problems over $\mathcal{R}$, cf. Fig. 6.1. Weighted sum rate maximization over $\mathcal{R}$ is a utility maximization problem, and a local solution can be computed using the natural global parameterization of $\operatorname{par}(\mathcal{R})$. Accordingly, efficient solutions are obtained by replacing a global maximizer of the weighted sum rate maximization problem by a local solution in the Lagrange duality based method. If the local solution is not globally optimal, an incorrect value and subgradient of the dual function result. As a consequence, the dual method may fail to converge, and if it converges, the resulting primal solution need neither be global nor locally optimal. However, simulation results show a good performance of this heuristic.

6.1.4 Numerical Results

Employing multi-antenna techniques, a base station can serve multiple users at the same time and at the same frequency. Accordingly, an optimum transmission strategy generally allows for interference between users. As evidenced by Sect. 6.1.2, interference leads to nonconvex optimization problems, and thereby to high complexity solution methods. In contrast, transmission schemes that are based on *orthogonal multiple access* (OMA) generally do not achieve optimal performance, but result in significantly simpler parameter optimization problems. A simple orthogonal multiple access scheme is given by a scheme that performs time sharing between the single-link points of the MISO channel. In this scheme, the physical layer parameters are the scheduling coefficients of the single-link points, and the rate for a given parameter setup is simply the convex combination of the single-link points for a particular choice of coefficients. For a concave utility u, the resulting parameter optimization problem is clearly convex (in fact, it is an instance of the "simple" utility maximization problem discussed in Sect. 3.1). Clearly, the increased complexity of optimum transmission strategies is only justified if the optimum strategies can provide a significant performance gain over the simple OMA strategy. In this section, simulation results are used to investigate the performance gains achievable by optimal strategies. In addition, further questions that are addresses is the performance of local solutions, and the gain achievable by scheduling for different system configurations.

Each choice of the channels $(\boldsymbol{h}_1, \ldots, \boldsymbol{h}_K)$ results in a realization of a utility maximization problem. In the simulations, the utility achieved by each of the solution methods is averaged over multiple such realizations. The channels $\boldsymbol{h}_k$ are drawn from a circularly symmetric complex Gaussian distribution, the channel coefficients are i.i.d. with unit power, and the users' channels are uncorrelated.

Only concave utility functions u are considered, resulting in problems of type 1 and 2, depending on whether scheduling is performed or not. In other words, nonconvexity occurs at the physical layer only. The opposite case of a convex physical layer model and a nonconvex utility model is investigated in the second case study in Sect. 6.2.

The following utility functions are simulated:

- Best effort:

$$u(\boldsymbol{s}) = \sum_{k=1}^{K} \ln(1 + s_k),$$

- Max–min:

$$u(\boldsymbol{s}) = \min_{k \in \{1,\dots,K\}} s_k,$$

- Sum log-rate:

$$u(\boldsymbol{s}) = \sum_{k=1}^{K} \ln(s_k).$$

The following solution methods are considered:

- Optimum precoder selection, no scheduling (*NoSc*), equivalent to maximizing utility over the rate region $\mathcal{R}$.
- Optimum scheduling and precoder selection (*OptSc*), equivalent to maximizing utility over the rate region $\mathcal{C}$.
- Precoder selection based on a local optimality criterion, no scheduling (*Local*).
- Precoder selection and scheduling using a local optimality criterion in the weighted sum rate maximization subproblems (*LocSc*).
- Orthogonal multiple access with optimum scheduling of single link points (*OMA*).

The local methods use Rosen's projected gradient algorithm [61] to obtain a local solution, starting at an initial power allocation

$$\boldsymbol{p}^{(0)} = \frac{P}{K}\mathbf{1}.$$

Figures 6.2–6.4 show the average utility achieved by a system with $N = 4$ antennas, $K = 3$ users, and different values of SNR $= \frac{P}{\sigma^2}$.

For all three utility functions under consideration, the performance of the methods NoSc, OptSc, Local, and LocSc is virtually identical. In contrast, in all three cases the performance of the orthogonal multiple access scheme is

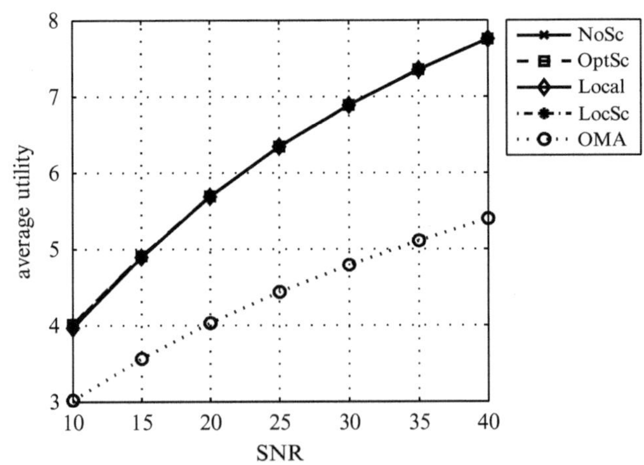

Fig. 6.2 Average utility, best effort, $N = 4$, $K = 3$

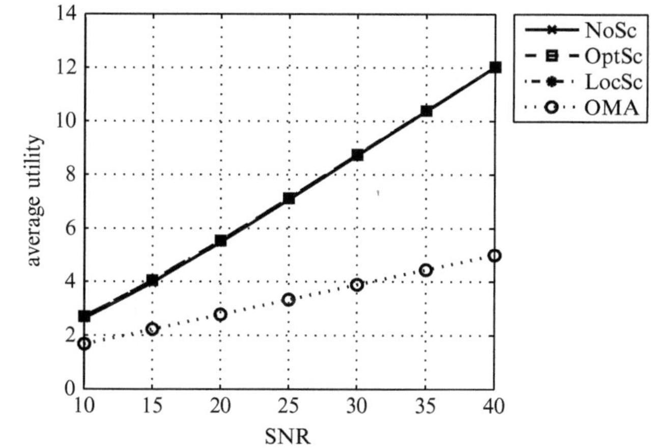

Fig. 6.3 Average utility, max–min, $N = 4$, $K = 3$

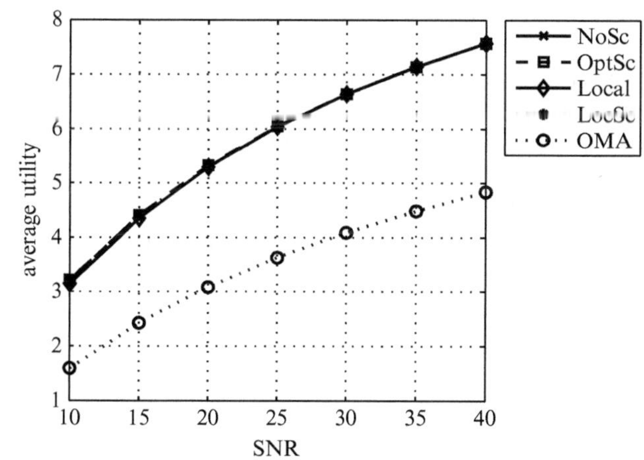

Fig. 6.4 Average utility, sum log-rate, $N = 4$, $K = 3$

significantly worse than that of the other methods. Clearly, it is highly beneficial to exploit the spatial degrees of freedom by serving multiple users in the same time slot.

Comparing the performance of NoSc and OptSc, it turns out that optimum scheduling does not provide a performance gain in the scenario under consideration. This result stems from the fact that the number of users in the system is smaller than the number of spatial degrees of freedom, which allows the base station to serve all users at the same time. As a result, it is not necessary to separate users via scheduling. Note that both max–min and sum log-rate utility enforce that all users get a non-zero rate (provided the rate region is proper). Accordingly, as the NoSc strategy cannot separate users via scheduling, all users are served at the same time by the NoSc strategy in Figs. 6.3 and 6.4.

It can also be observed that the local methods achieve a performance that is practically optimal. Thus, while global optimality cannot be guaranteed, the observed performance of local methods that operate on the Pareto boundary of the rate region is indeed very good. Taking into account that the computational complexity of local methods is significantly lower, it can be concluded that local methods provide an excellent performance-complexity trade-off.

For the results shown in Figs. 6.2–6.4, the number of users served by the base station is increased to $K = 4$. A slight gain by optimum scheduling is noticeable. However, this slight gain does hardly justify the significantly higher computational effort required to find the optimum scheduling solution. Performance of the local methods is again virtually optimal, while the OMA method again suffers from a significant performance loss, due to its inability to serve more than one user at the same time.

Based on the results from Figs. 6.2–6.7 an obvious question to ask is whether scheduling is necessary at all. An answer is provided by Figs. 6.8–6.10.

For the result in Figs. 6.8–6.10, the number of transmit antennas at the base station is reduced to $N = 2$, while the number of users is still $K = 4$. Due to the insufficient number of spatial degrees of freedom, it is clearly advantageous to not serve all users at the same time. Instead, scheduling between $K = 4$ rate vectors provides a significant gain in performance. The gains achieved by scheduling are most significant in the case of max–min and sum log-rate utility, as both utility functions enforce non-zero rate for all users. Without scheduling, this requires the base station to serve all $K = 4$ users at the same time, with only two transmit antennas. The resulting strong interference between users leads to a severe reduction in terms of achievable system utility. The performance loss is less severe in case of a best effort utility, as for this utility the base station can decide to not serve some of the users, a strategy that is clearly beneficial in case of insufficient degrees

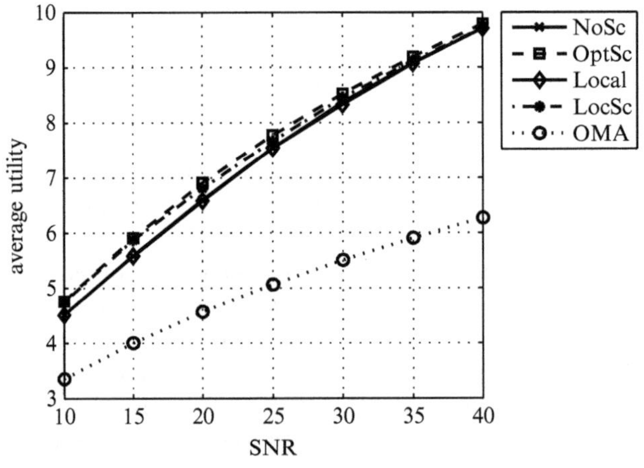

Fig. 6.5 Average utility, best effort, $N = 4$, $K = 4$

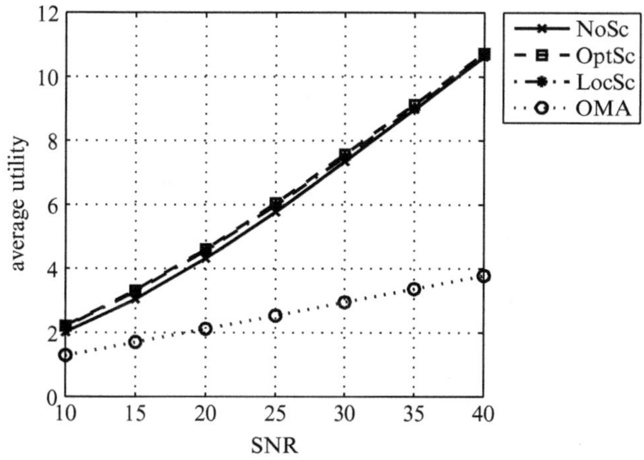

Fig. 6.6 Average utility, max–min, $N = 4$, $K = 4$

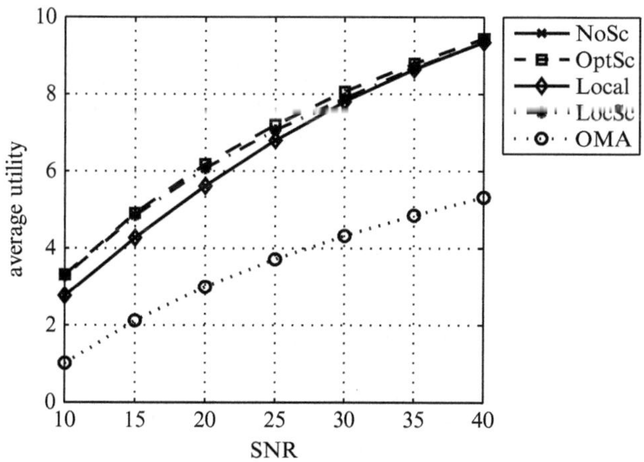

Fig. 6.7 Average utility, sum log-rate, $N = 4$, $K = 4$

of freedom. Remarkably, OMA outperforms NoSc for all three utility models. In other words, scheduling of single user points provides larger gains than interference management by beamforming. Still, by far the best performance is achieved by jointly optimal scheduling and beamforming. Comparing the performance of OptSc and OMA, it is evident that again a significant gain is achieved by serving more than a single user at the same time. Note that the OptSc strategy implicitly solves the user selection problem in an optimal manner—which users to serve follows directly from the optimal parameter setup. In other words, the optimal user selection is a *result* of the optimization.

With respect to the performance of local methods, it can be observed that they again perform close to optimal. The performance of the local method without

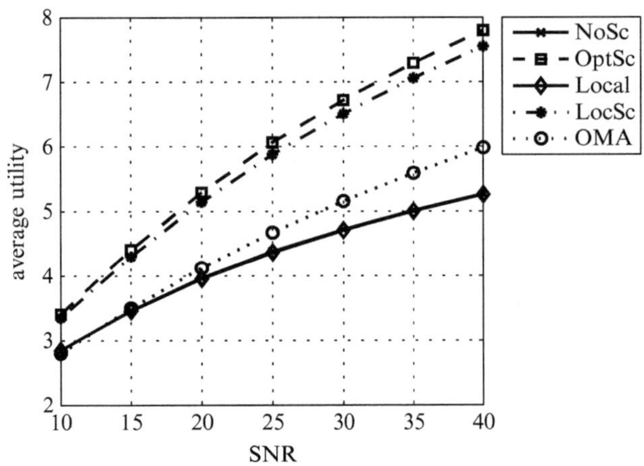

Fig. 6.8 Average utility, best effort, $N = 2$, $K = 4$

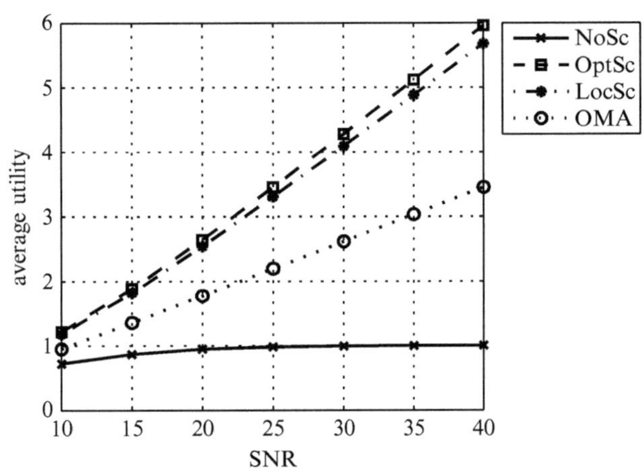

Fig. 6.9 Average utility, max–min, $N = 2$, $K = 4$

scheduling is upper bounded by the performance of the NoSc strategy. While this upper bound is attained for a majority of channel realization, the absence of a scheduling mode again results in rather poor performance. In contrast, the combination of a Lagrange duality based approach to solving the rate space problem on $\mathcal{C}$ and a local method for (approximately) solving the weighted sum rate maximization problems yields close to optimal performance at feasible computational complexity. Hence, the LocSc strategy can be identified as the most promising strategy for systems serving a larger number of users, where the global optimal strategy OptSc is no longer practically feasible.

Figures 6.11–6.13 show the performance of the different strategies for a varying number of users K, $N = 4$ transmit antennas at the base station, and a fixed

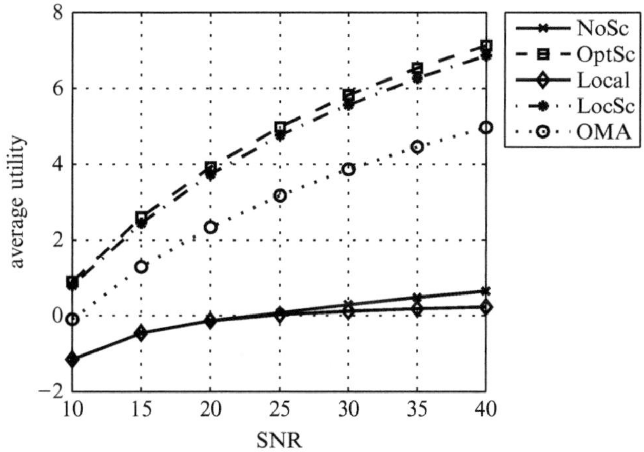

Fig. 6.10 Average utility, sum log-rate, $N = 2$, $K = 4$

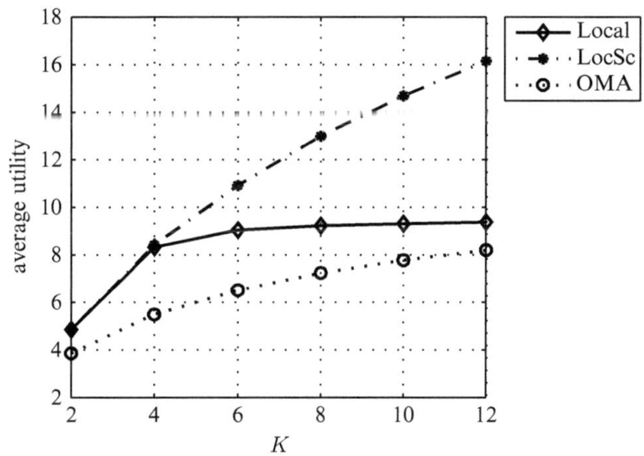

Fig. 6.11 Average utility, best effort, $N = 4$, SNR = 30 dB

SNR of 30 dB. Due to their prohibitively high complexity, the globally optimal methods OptSc and NoSc are not included in Figs. 6.11–6.13. Only in case of a max–min utility is computing the globally optimal solution without scheduling (NoSc) practically feasible, as discussed before. In all three cases, scheduling provides significant gains as soon as the number of users exceeds the number of transmit antennas.

In the case of a best-effort utility, system utility increases with the number of users. This result is due to the fact that with a best effort utility, the system can decide to only service a subset of users. In other words, the base station can pick the users that provide best overall performance. In case of OMA, this corresponds to choosing the users with the best channels, and the more users there are in the system,

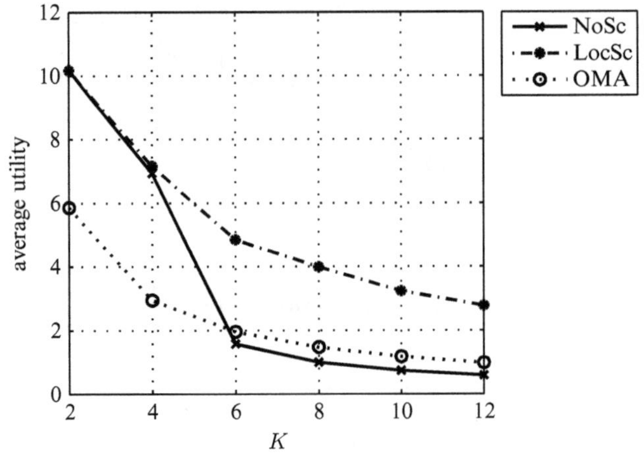

Fig. 6.12 Average utility, max–min, $N = 4$, SNR $= 30$ dB

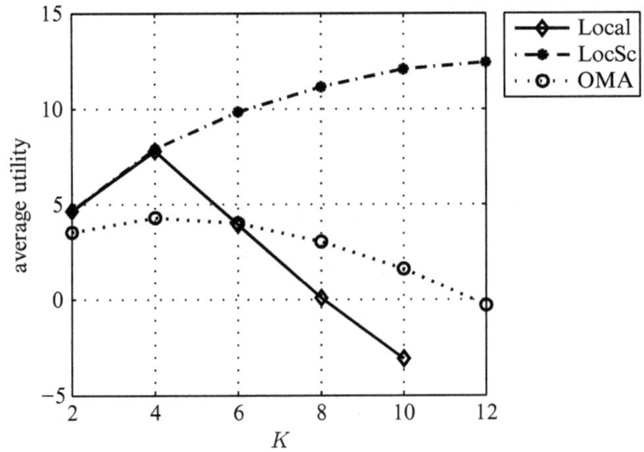

Fig. 6.13 Average utility, sum log-rate, $N = 4$, SNR $= 30$ dB

the more likely it is to find a subset of users with good channels. The resulting increase in performance is also known as *multi user diversity* [84]. In contrast, for a max–min utility, system performance decreases with the number of users. This result is an immediate consequence of the fact that max–min results in the same rate for all users.

Considering Fig. 6.12, it can be observed that performance of the NoSc solution is drastically reduced as soon as the number of users exceeds the number of transmit antennas. As a result, for a max–min utility, the NoSc solution is outperformed by the simple OMA strategy for $K \geq 6$.

Figure 6.13 shows that a similar behavior results for the sum log-rate utility. Again, as long as $K \leq N$, the gain of scheduling is negligible, while for $K > N$, the performance of the local method without scheduling is drastically reduced. Notably, for the values of K under consideration, the achieved average utility of the LocSc approach increases with the number of users, while the performance of both LocSc and OMA is maximal at $K = N = 4$ and then decreases with K. As a result, for $K \geq 6$, the LocSc strategy outperforms the other two strategies by a wide margin. The results of the Local method in Fig. 6.13 were obtained by using an SQP/Quasi-Newton [25] solver for computing a stationary point. For $K \geq 6$, a purely gradient-based approach frequently exhibited a zig-zagging behavior, which caused extremely slow convergence.

To summarize the numerical results for concave utility functions, in order to achieve optimum performance, a joint optimization of beamformers and scheduling is clearly required if the number of users exceeds the number of transmit antennas. For all three utility functions under consideration, beamforming without scheduling only represents a good strategy for $K \leq N$. In contrast, orthogonal multiple access with optimum scheduling outperforms optimum beamforming for $K > N$ and max–min or sum log-rate utility, but cannot match the performance of joint beamforming and scheduling, regardless of the utility function under consideration. The observed performance of the local methods is close to optimum. As a conclusion, local methods that operate on the Pareto boundary of the rate region offer an excellent performance-complexity trade-off. Evidently, drawing this conclusion is only possible *after* establishing the ultimate performance limits by global methods.

6.2 Nonconcave Utility Maximization in the MIMO Broadcast Channel

In the second case study, utility maximization in a MIMO broadcast channel with *dirty paper coding* (DPC) at the transmitter and time sharing is investigated. A detailed description of the corresponding physical layer model is provided in Sect. 4.2. The capacity region $\mathcal{C}$ of the MIMO broadcast channel is convex and proper (under mild assumptions, cf. Proposition 4.2.2).

Concerning the utility model (u, h), two different classes of utility functions u are considered: concave and nonconcave. There are no minimum utility constraints, i.e., $h(\mathbf{0}) = 0$.

The parameter space problem in the MIMO BC is generally nonconvex, regardless whether the utility function is concave or not, see Sect. 6.2.1. In contrast, combining a concave utility model with a convex capacity region leads to a convex rate space problem. As discussed in Sect. 6.2.2, convex rate space problems in a broadcast channel with dirty paper coding can be efficiently solved by Lagrange duality. In contrast, nonconcave utility functions lead to a nonconvex rate space problem. Still, by exploiting the monotonic structure of the rate space problem, it is possible to solve the nonconcave utility maximization problem to global optimality. Again, however, application of global methods is only practically feasible for a relatively small number of users K. Hence, in case of a nonconcave utility functions, local methods provide good solutions at practically feasible complexity.

6.2.1 Parameter Space Problem

For a given utility function u, the parameter space problem in the MIMO broadcast channel with DPC and time sharing is given by

$$\max_{\boldsymbol{x}_1'} u\left(\sum_{k=1}^{K} b_k \boldsymbol{r}(\boldsymbol{Q}^k, \pi^k)\right) \quad \text{s.t.} \quad \boldsymbol{x}_1' \in \mathcal{X}_1',$$

where

$$\boldsymbol{x}_1' = (\boldsymbol{b}, \boldsymbol{Q}^1, \pi^1, \ldots, \boldsymbol{Q}^K, \pi^K),$$
$$\mathcal{X}_1' = \left\{\boldsymbol{x}_1' : \boldsymbol{b} \in \mathbb{R}_+^K, \mathbf{1}^{\mathrm{T}}\boldsymbol{b} = 1, (\boldsymbol{Q}^k, \pi^k) \in \mathcal{Q} \times \Pi, k = 1, \ldots, K\right\}.$$

Due to the nonconcavity of the rate map $\boldsymbol{r}$ and the discrete nature of the set of possible encoding orders Π, the parameter space problem is nonconvex.

Concerning the structure of the parameter space problem, the same statement as in case of a MISO broadcast channel with linear precoding applies: the parameter space problem is a nonconvex optimization problem with no structure except for continuity.

Weighted sum rate maximization represents a special case. For $u(\boldsymbol{s}) = \boldsymbol{\lambda}^{\mathrm{T}}\boldsymbol{s}$, with $\boldsymbol{\lambda} \geq \mathbf{0}$, the rate space formulation is nonconvex as well. Still, in this special case, time sharing is not necessary and it is possible to infer the optimal encoding order from the weight vector $\boldsymbol{\lambda}$ [77]. Finally, by using an uplink-downlink duality [76], it is possible to obtain a convex parameter space formulation in the dual MAC. Hence, in the special case of weighted sum rate maximization, it is possible to directly compute an optimal parameter setup by efficient algorithms [77, 79, 80].

6.2.2 Globally Optimal Solutions

Globally optimal solutions can be obtained based on the rate space formulation introduced in Sect. 3.2. Let $\mathcal{C}$ denote the capacity region of the MIMO BC as defined in Sect. 4.2. The rate space problem is given by

$$\max_{\boldsymbol{s}} u(\boldsymbol{s}) \quad \text{s.t.} \quad \boldsymbol{s} \in \mathcal{C}. \tag{6.5}$$

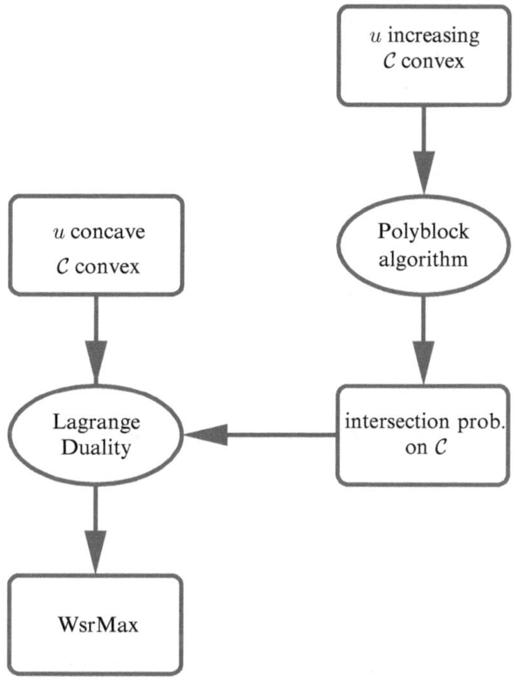

Fig. 6.14 Globally optimal solution methods in the MIMO BC

The capacity region $\mathcal{C}$ is convex. Hence, if u is concave, then Problem (6.5) is a convex problem. Moreover, as discussed in Sect. 4.2, in the MIMO BC with dirty paper coding, weighted sum rate maximization is efficiently solvable. As a result, for concave utility functions, Problem (6.5) can be solved by Lagrange duality, by following the steps described in Sect. 3.3.

In case of a nonconcave utility function u, the extended polyblock algorithm from Sect. 3.4.3 can be used to compute an optimal rate vector. At each iteration of the polyblock algorithm, an intersection problem over $\mathcal{C}$ has to be solved. Due to the fact that $\mathcal{C}$ is proper, an intersection problem corresponds to a generalized max–min problem. Due to the convexity of $\mathcal{C}$, generalized max–min results in a convex rate space problem, which, as discussed before, can be solved by Lagrange duality. The modular structure of the global solution methods is summarized in Fig. 6.14. Evidently, the elementary building block for both methods is weighted sum rate maximization. Consequently, the same set of solution methods can be applied to any physical layer for which weighted sum rate maximization is efficiently solvable.

6.2.3 Local Solutions

A local solution of the rate space problem (6.5) can be computed by using the methods from Sect. 3.5. The local method uses a global parameterization of the

Pareto boundary of $\mathcal{C}$. In contrast to the MISO broadcast channel with linear precoding, a closed-form parameterization is not available. Instead, the generic parameterization introduced in Sect. 3.5.3 is used, as discussed in Sect. 4.2.4. In Sect. 4.2.4 it is also shown that local methods are only useful in case of nonconcave utilities. For concave utilities, evaluating the parameterization at a single point is of similar complexity as computing the optimum rate vector.

6.2.4 Numerical Results

For concave utilities, the optimum rate vector can be computed by Lagrange duality. Due to the convexity of the rate space problem and the availability of efficient algorithms for maximizing a weighted sum of rates, it is not necessary to apply global methods. Moreover, while the optimum rate could also be computed by a local method, such an approach is much less efficient than the dual method. The situation is entirely different for nonconcave utilities. In this case, it is generally no longer possible to compute the optimum rate vector by Lagrange duality. In order to find a global maximizer, it is necessary to use the global method from Sect. 3.4. Moreover, due to the complexity of global methods, local methods become an attractive choice. In the following, the performance of the different solution methods is investigated for a nonconcave utility function.

The utility function u is defined as the point-wise sum[1] of the users' utilities,

$$u(\boldsymbol{s}) = \sum_{k=1}^{K} u_k(s_k). \tag{6.6}$$

For each user, the sigmoidal utility function u_k defined in (5.5) is used. Due to the nonconcavity of the sigmoidal utilities, the utility function u is nonconcave.[2] As a result, Lagrange duality based approaches cannot be used to compute the optimal rate vector. Still, due to the additive structure of the utility function u, the dual subproblem

$$\max_{\boldsymbol{z} \in \mathcal{Z}} u(\boldsymbol{z}) - \boldsymbol{\lambda}^{\mathrm{T}} \boldsymbol{z} \tag{6.7}$$

decouples into K one-dimensional problems

[1]The term "point-wise sum" is used to distinguish the sum in (6.6) from the usual definition of a sum of functions.

[2]A slightly weaker property than concavity is quasi-concavity [28]. If a cost function is quasi-concave, the convexity of its level sets can be exploited to find efficient solutions [28]. While every increasing function with domain $\mathbb{R}_+$ is quasi-concave, the sum of quasi-concave functions is generally not quasi-concave. In particular, a point-wise sum of sigmoidal functions as defined in (6.6) is not quasi-concave.

$$\max_{0 \le z_k \le s_k^{\mathrm{U}}} u(z_k) - \lambda_k z_k. \tag{6.8}$$

By evaluating the KKT conditions of Problem (6.8), it turns out that the optimum of Problem (6.8) is attained at either $z_k = 0, z_k = s_k^{\mathrm{U}}$, or one of the points where

$$\nabla u_k(z_k) = \lambda_k. \tag{6.9}$$

Finally, due to the fact that ∇u_k is strictly increasing for $s_k < b_k$ and strictly decreasing for $s_k \ge b_k$, there exist at most two points satisfying (6.9), and both can be found by bisection. Hence, Problem (6.8) (and thereby the dual subproblem (6.7)) can be solved to global optimality at low complexity.[3] As shown in Sect. 3.3.5, if the dual subproblem can be solved, Lagrange duality can be used to compute an upper bound for the maximally achievable utility value and a feasible point $\hat{\boldsymbol{s}}$. The point $\hat{\boldsymbol{s}}$ is the globally optimal rate vector if the utility function is replaced by its concave envelope. Let $\tilde{u}$ denote the concave envelope of u, with u defined by (6.6). Then

$$\tilde{u}(\boldsymbol{s}) = \sum_{k=1}^{K} \tilde{u}_k(s_k),$$

where $\tilde{u}_k$ is the concave envelope of u_k. For a sigmoidal utility function, u_k and $\tilde{u}_k$ coincide at $s_k = 0$ and for sufficiently large s_k, as shown by Fig. 5.1. As a result, it can be expected that the dual upper bound is tight and the utility achieved by $\hat{\boldsymbol{s}}$ close to the optimal solution for high SNR and sufficient degrees of freedom (as in this case, the rate provisioned to each user is sufficiently large), as well as for very low SNR, as the difference between u_k and $\tilde{u}_k$ is small for rates close to zero. Figure 5.1 also shows that the maximal difference between u_k and its concave envelope decreases with a. Hence, the performance of a Lagrange duality based approach is worst for large a and medium SNR, as in this case $\tilde{u}_k$ significantly over-estimates u_k in the range of relevant rates.

Figure 6.15 shows the average utility achieved by different optimization strategies for a system with $N = 6$ transmit antennas and $K = 3$ receivers. Each receiver is equipped with $M_k = 2$ receive antennas. For each user, the channel coefficients are drawn from a circularly symmetric Gaussian distribution with zero mean and unit variance. The channel coefficients are uncorrelated across antennas and users. Results are averaged over 1,000 channel realizations.

[3]In the model under consideration, the utility function and its gradient are available in closed-form. Hence, evaluating the gradient is possible at a low computational cost. In a real-world system, such a closed-form description of the upper layers may not be available. In this case, it is clearly possible that evaluating the utility function (or its gradient) is a computationally expensive operation. Throughout this work, however, the cost of evaluating the utility function (and its (sub-) gradient) is not taken into account.

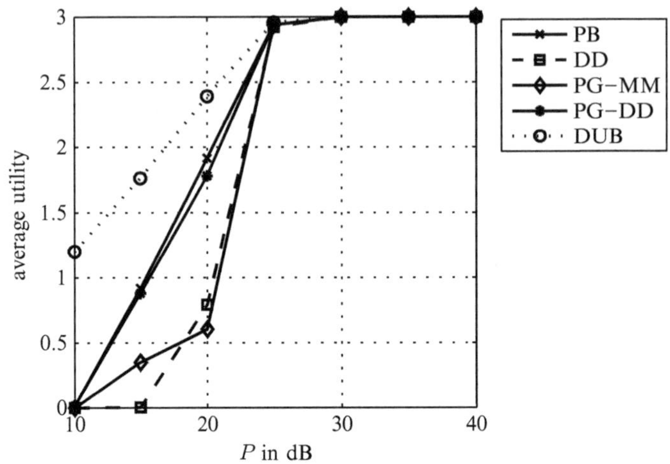

Fig. 6.15 Average utility, $a = 2.5$, $N = 6$, $K = 3$, $M_k = 2$

The graph with label DD shows the average utility achieved by the rate vector $\hat{s}$ obtained by applying Lagrange duality to solve the nonconvex rate space problem. Furthermore, the graph DUB shows the average utility value of the dual upper bound. The graphs PG-DD and PG-MM correspond to local solutions. For both PG-DD and PG-MM, Rosen's projected gradient method is used to find a stationary point on the Pareto boundary of $\mathcal{C}$. Stepsizes are determined by an inexact line-search similar to Armijo's method. While PG-DD uses as initial point the rate vector obtained by the dual approach, PG-MM starts at the max–min rate vector. The projected gradient uses two termination criteria: it terminates if either the angle between gradient and tangent space is larger than 89° or if the norm of the projected gradient is smaller than 10^{-6}. The graph PB shows the average utility achieved by the polyblock algorithm. The termination criterion of the polyblock algorithm is ϵ-optimality of the utility value of the best feasible rate point, with ϵ chosen as 0.1. The average utility plotted in Fig. 6.15 is the average of the utility of the best feasible rate vector. The polyblock algorithm starts with the rate point computed by PG-DD as the current best feasible solution.

As expected, the dual method provides optimal performance at high SNR. In contrast, in the medium SNR range, the rate vector obtained by applying Lagrange duality is far from optimal. Moreover, the dual upper bound is rather loose for low to medium SNR and gets tight at high SNR. The local method PG-DD provides close to optimal performance for all values of transmit power P. In the medium SNR range, a significant improvement over the initial point is achieved. In contrast, the performance of the PG-MM method is rather poor. This result can be explained at follows: At the max–min point, all users have equal rate. As the utility function u_k is the same for all users, the gradient of u at the max–min point is a scaled version of the all-ones vector. Accordingly, if the max–min point is (close to) a sum-rate maximizer, tangent space of $\mathcal{C}$ and gradient of u at the max–min point are (close to) orthogonal—i.e., in this case the max–min point is a stationary point of the rate space problem and the local method terminates. The max–min point is close to a sum-rate maximizer if the capacity region $\mathcal{C}$ is approximately symmetric. Due to the fact that the users' channel coefficients are drawn from identical distributions, such an approximate symmetry seems to happen quite frequently. A comparison of PG-DD and PG-MM shows the impact of the choice of initial point on the performance of local methods.

For $P \geq 25$ dB, all three users are provided with the maximum possible utility value, and system performance saturates. In other words, a value of $P = 25$ dB is sufficient to serve three users. Higher sum-utility can only be obtained by adding more users to the system. For the results shown in Fig. 6.16, the number of users is increased to $K = 5$, while all other system parameters are left unchanged. Again, for sufficiently high SNR, all methods provide optimum performance. In the medium SNR range, however, increasing the number of users *impairs* the performance of all but the globally optimum method. This result can be explained on the basis of Figs. 5.1 and 6.17.

Figure 5.1 shows the difference between the nonconcave utility function u_k and its concave envelope, while Fig. 6.17 shows the average number of users that is served with a rate greater than $b = 12$. The graph of the polyblock method in Fig. 6.17 shows that at low and medium SNR, the optimal strategy is to focus resources on a subset of users. The concave envelope severely overestimates the actual utility value for small rates. As a consequence, the dual method tends to share resources among too many users, and the number of users that receive a rate greater than b is too small, as shown by Fig. 6.17. Determining the optimal subset of users also represents a problem for the PG-DD method. Recall that the initial point is provided by the dual method, which tends to share resources among too

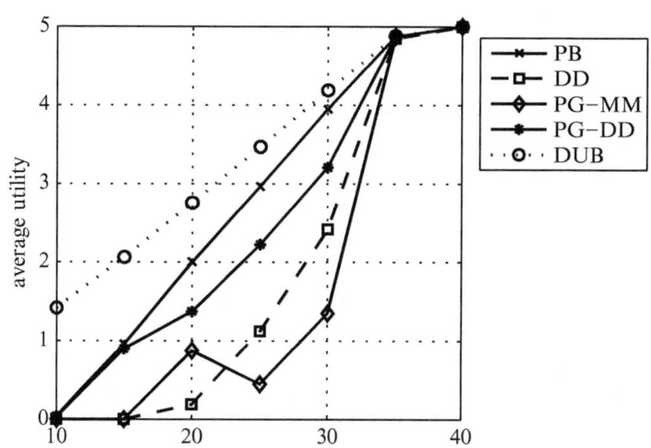

Fig. 6.16 Average utility, $a = 2.5$, $N = 6$, $K = 5$, $M_k = 2$

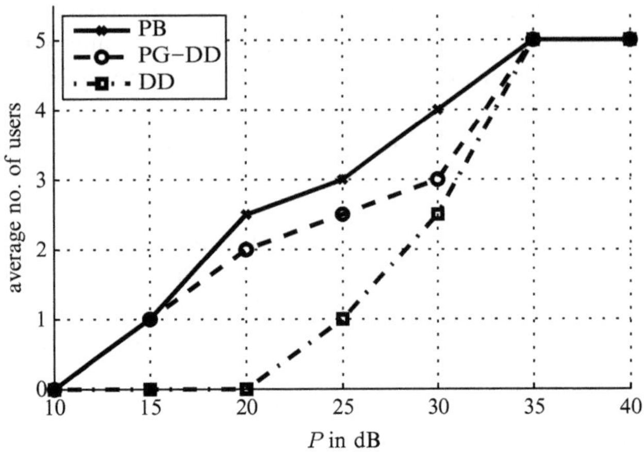

Fig. 6.17 Average number of users with $s_k > b$

many users. If the rate assigned to a certain user is too small at the initial point, the local information that is available does not indicate that utility can be maximized by increasing the rate of this particular user. From Fig. 5.1, it can be concluded that for large a, a gradient-based algorithm tries to increase the rates of the users whose current rate is close to the inflection point b. For medium SNR, the subset of users whose initial rate is close to or larger than the inflection point may be smaller than the subset of users served at the global optimum. As a result, starting from a point that shares resources among too many users, the PG-DD method tends to focus resources on too few users. Still, the PG-DD method can again achieve a significant gain over the dual method. In contrast, the PG-MM method provides the worst performance among all methods under consideration.

The convergence rate of the PG-DD method is examined in Fig. 6.18. Average utility is plotted versus the number of gradient iterations. Each graph in Fig. 6.18 corresponds to a value of transmit power P, where P takes the same values as in Fig. 6.16. Hence, the rightmost point of each graph in Fig. 6.18 corresponds to an average utility value in Fig. 6.16. Not shown is the graph for $P = 40\,\text{dB}$, as for this value of P the dual method already succeeds in computing the optimal solution, hence the PG-DD method terminates at the first iteration. Figure 6.16 shows that a significant improvement over the initial utility value is achieved after 5 gradient iterations. Moreover, between 10 and 20 gradient iterations are sufficient to find a local solution.

The convergence behavior of the polyblock algorithm is shown in Fig. 6.19. Average utility is plotted versus the number of polyblock iterations. For each value of P, two graphs are shown (solid and dashed line), and $P \in \{10\,\text{dB}, 15\,\text{dB}, 20\,\text{dB}, 25\,\text{dB}, 30\,\text{dB}\}$, resulting in 5 pairs of graphs. For each pair, the solid line corresponds to the lower bound given by the best utility value obtained from a feasible rate vector, while the dashed line corresponds to the upper bound

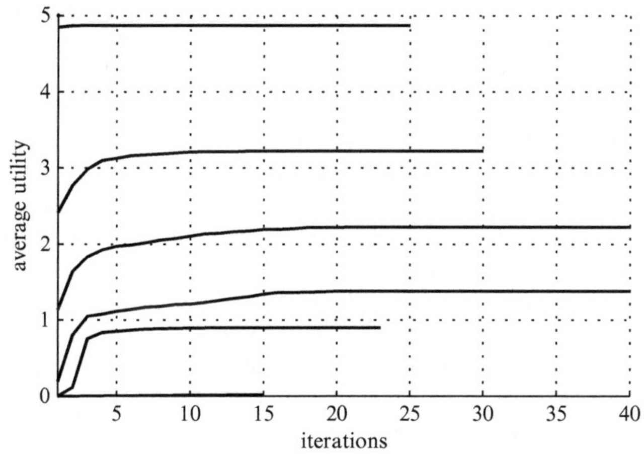

Fig. 6.18 Convergence behavior, PG-DD, $a = 2.5$, $N = 6$, $K = 5$, $M_k = 2$

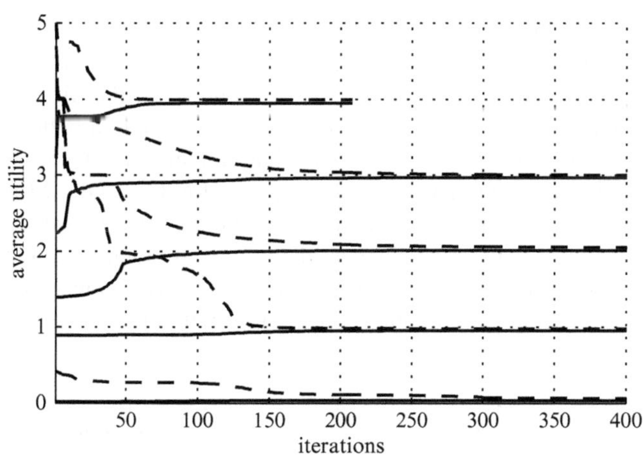

Fig. 6.19 Convergence behavior, PB, $a = 2.5$, $N = 6$, $K = 5$, $M_k = 2$

given by the utility value of the best vertex of the approximating polyblock, cf. Sect. 3.4.1.

It can be observed that the lower bound is close to the global optimum after at most 50 iterations. However, except for $P = 30$ dB, it takes between 150 and 200 iterations until a certificate of global optimality is obtained, due to the rather slow convergence of the upper bound. Recall that the polyblock algorithm is initialized with the rate vector obtained by the PG-DD method. For small values of P, the rate vector computed by the PG-DD method is close to optimal. Still, it takes the polyblock algorithm 150 iterations to verify this optimality. On the other hand, for $P = 30$ dB, a significant improvement over the suboptimal initial value is achieved after only a few polyblock iterations.

Figure 6.19 exhibits the average behavior of the polyblock algorithm. Also of interest is the observed worst-case complexity of the algorithm. Figure 6.20 shows the average, medium, and maximum number of iterations until the desired precision $\epsilon = 0.1$ is reached, for $K = 5$ users. The maximum number of iterations that was observed in 1,000 simulation runs is approximately 1,500. While the computational complexity is much higher than that of the suboptimal methods, it is still relatively low, considering that a global method is used. For $a = 2.5$, the users' utility functions make a steep transition between minimal and maximal value in a small interval around rate b, and are almost flat outside of this interval. This almost binary behavior is beneficial for the polyblock algorithm: the quick decay to zero below rate b helps the algorithm to identify regions that are not of interest. Intuitively speaking, there is a quick transition between areas of optimal and suboptimal points, hence a rough approximation of $\mathcal{C}$ is sufficient in suboptimal regions. Moreover, the saturation for values larger than b leads to good upper bounds, even if the underlying rate vector is not close to a maximizer.

Numerical results in Figs. 6.15 and 6.16 are based on a sigmoidal utility function with transition parameter $a = 2.5$. As shown in Fig. 5.1, such a choice of a corresponds to a rather steep transition between minimal and maximal value. For a smaller value of a, a smoother transition results. Hence, for a smaller value of a, the following results can be expected: First, for smaller a, the difference between utility function and concave envelope is reduced. As a result, a better performance of the dual method as well as a tighter dual upper bound can be expected. The improved initial value also helps the local method PG-DD. Moreover, for a smoother transition, the range in which the gradient provides useful information is enlarged. Hence, a good performance of the PG-DD method can be expected. With respect to the polyblock algorithm, a smoother transition leads to larger areas of good, but suboptimal points, and a finer approximation of the rate region is

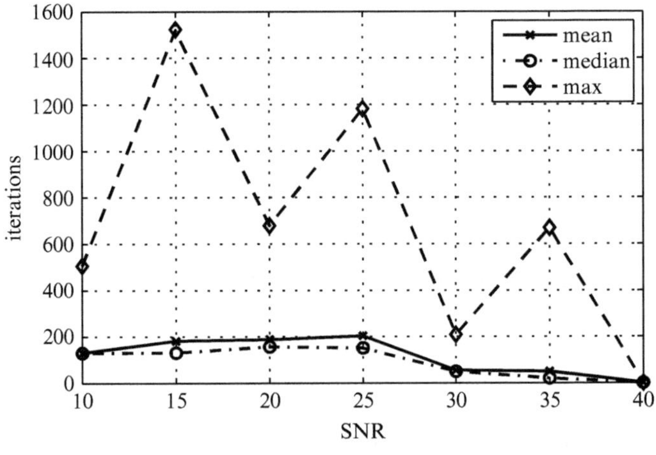

Fig. 6.20 Polyblock iterations, $a = 2.5$, $N = 6$, $K = 5$, $M_k = 2$

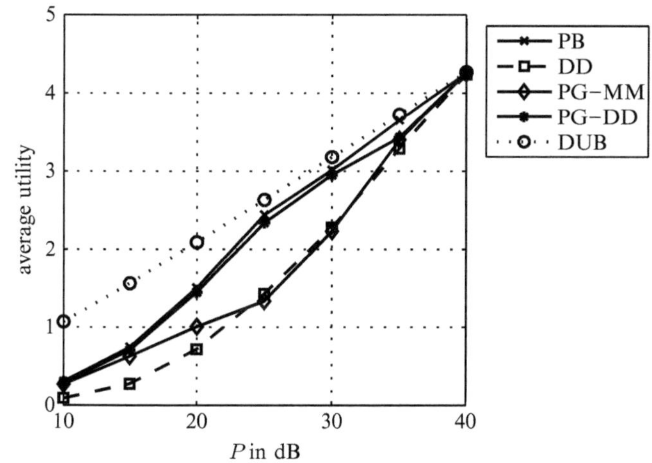

Fig. 6.21 Average utility, $a = 0.5$, $N = 6$, $K = 5$, $M_k = 2$

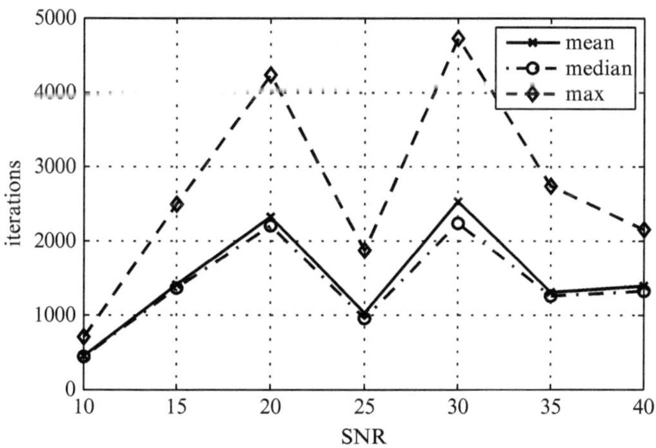

Fig. 6.22 Polyblock iterations, $a = 0.5$, $N = 6$, $K = 5$, $M_k = 2$

required. Accordingly, it can be expected that computational complexity of the polyblock algorithm increases with decreasing a. Figure 6.21 shows the average utility achieved by the different methods for $a = 0.5$ and $K = 5$ users.

As expected, performance of the PG-DD and the dual method is improved relative to the global optimum. Finally, Fig. 6.22 investigates the number of iterations needed by the polyblock algorithm to achieve an accuracy of $\epsilon = 0.1$. As expected, decreasing a leads to an increase in the number of iterations needed by the polyblock algorithm.

Appendix A

A.1 Mathematical Preliminaries

A.1.1 Sets

Let $\mathcal{X}$ and $\mathcal{Y}$ be two arbitrary sets. Then

$$
\begin{aligned}
&\mathcal{X} \subseteq \mathcal{Y} \iff (x \in \mathcal{X} \implies x \in \mathcal{Y}), \\
&\mathcal{X} \subset \mathcal{Y} \iff (\mathcal{X} \subseteq \mathcal{Y} \quad \text{and} \quad \mathcal{X} \neq \mathcal{Y}), \\
&\mathcal{X} \setminus \mathcal{Y} = \{x \in \mathcal{X} : x \notin \mathcal{Y}\}.
\end{aligned}
$$

A.1.2 Spaces

Let $\mathcal{V}$ be a vector space. Elements of $\mathcal{V}$ are denoted by bold letters, e.g., $\boldsymbol{x} \in \mathcal{V}$. In addition, functions that map into a vector space are also denoted by bold letters, e.g., $\boldsymbol{f} : \mathcal{X} \to \mathcal{V}$, where $\mathcal{X}$ is arbitrary.

The set of real numbers is denoted by $\mathbb{R}$. Moreover,

$$
\begin{aligned}
\mathbb{R}_{+} &= \{x \in \mathbb{R} : x \geq 0\}, \\
\mathbb{R}_{++} &= \{x \in \mathbb{R} : x > 0\}.
\end{aligned}
$$

The set of complex numbers is denoted by $\mathbb{C}$.

The Cartesian product of two sets $\mathcal{X}$ and $\mathcal{Y}$ is denoted by $\mathcal{X} \times \mathcal{Y}$, and $\mathcal{K}^N$ is the N-fold Cartesian product of $\mathcal{K}$. An exception is made if $\mathcal{K} \in \{\mathbb{R}, \mathbb{C}\}$: $\mathbb{R}^K$ denotes the K-fold Cartesian product of $\mathbb{R}$, equipped with the usual vector space structure, an inner product

J. Brehmer, *Utility Maximization in Nonconvex Wireless Systems*, Foundations in Signal Processing, Communications and Networking 5, DOI 10.1007/978-3-642-17438-4,

$$\langle \boldsymbol{x}, \boldsymbol{y} \rangle = \sum_{k=1}^{K} x_k y_k,$$

the induced norm

$$\|\boldsymbol{x}\|_2 = \sqrt{\langle \boldsymbol{x}, \boldsymbol{x} \rangle},$$

the metric

$$d(\boldsymbol{x}, \boldsymbol{y}) = \|\boldsymbol{x} - \boldsymbol{y}\|_2,$$

and the standard topology induced by the metric d. Likewise, $\mathbb{C}^K$ is the K-fold Cartesian product of $\mathbb{C}$, equipped with the usual vector space structure, inner product, norm, and metric.

The vector $\mathbf{1} \in \mathbb{R}^K$ is the vector of all-ones, and $\mathbf{0} \in \mathbb{R}^K$ is the vector of all-zeros. Given $\boldsymbol{x} \in \mathbb{R}^K$ and $\eta \in \mathbb{R}_{++}$, an open ball $\mathcal{B}_\eta(\boldsymbol{x})$ is defined by

$$\mathcal{B}_\eta(\boldsymbol{x}) = \left\{\boldsymbol{y} \in \mathbb{R}^K : d(\boldsymbol{x}, \boldsymbol{y}) < \eta\right\}.$$

The interior of a set $\mathcal{X} \subseteq \mathbb{R}^K$ is

$$\operatorname{int}(\mathcal{X}) = \left\{\boldsymbol{x} \in \mathcal{X} : \exists \eta > 0 : \mathcal{B}_\eta(\boldsymbol{x}) \subset \mathcal{X}\right\}.$$

The boundary of a set $\mathcal{X} \subseteq \mathbb{R}^K$ is

$$\operatorname{bd}(\mathcal{X}) = \mathcal{X} \setminus \operatorname{int}(\mathcal{X}).$$

For $\mathcal{K} \in \{\mathbb{R}, \mathbb{C}\}$, the vector space of matrices with M rows and N columns and entries in $\mathcal{K}$ is denoted by $\mathcal{K}^{M \times N}$. Matrices with $M, N > 1$ are frequently denoted by uppercase letters, although this convention is not strictly adhered to. The matrix $\boldsymbol{X}^{\mathrm{T}}$ is the transpose of $\boldsymbol{X}$, $\boldsymbol{X}^{\mathrm{H}}$ is the conjugate transpose of $\boldsymbol{X}$, $\operatorname{rank}(\boldsymbol{X})$ is the rank of $\boldsymbol{X}$, and $\operatorname{tr}(\boldsymbol{X})$ denotes the trace of $\boldsymbol{X}$. The matrix $\boldsymbol{X}^{-1}$ is the inverse of $\boldsymbol{X}$, provided it exists. The identity matrix is $\boldsymbol{I}$, and $\boldsymbol{e}_n \in \mathbb{R}^{M \times 1}$ is the n-th column of $\boldsymbol{I} \in \mathbb{R}^{M \times N}$. Range and nullspace of a matrix $\boldsymbol{A} \in \mathbb{R}^{M \times N}$ are defined as follows:

$$\operatorname{range}(\boldsymbol{A}) = \left\{\boldsymbol{A}\boldsymbol{x} : \boldsymbol{x} \in \mathbb{R}^{N \times 1}\right\},$$
$$\operatorname{null}(\boldsymbol{A}) = \left\{\boldsymbol{x} \in \mathbb{R}^{N \times 1} : \boldsymbol{A}\boldsymbol{x} = \mathbf{0}\right\}.$$

By equipping $\mathbb{R}^{M \times N}$ with the inner product

$$\langle \boldsymbol{X}, \boldsymbol{Y} \rangle = \operatorname{tr}\left(\boldsymbol{X}^{\mathrm{T}} \boldsymbol{Y}\right)$$

and the metric induced by this inner product, $\mathbb{R}^{M \times N}$ and $\mathbb{R}^{MN}$ are isometric. Accordingly, no distinction is made between $\mathbb{R}^{K \times 1}$ and $\mathbb{R}^K$, and the inner product of $\boldsymbol{x}, \boldsymbol{y} \in \mathbb{R}^K$ is written as

$$\langle \boldsymbol{x}, \boldsymbol{y} \rangle = \operatorname{tr}\left(\boldsymbol{x}^{\mathrm{T}} \boldsymbol{y}\right) = \boldsymbol{x}^{\mathrm{T}} \boldsymbol{y}.$$

For two vectors $\boldsymbol{x}, \boldsymbol{y} \in \mathbb{R}^K$, the element-wise product $\boldsymbol{x}\boldsymbol{y} \in \mathbb{R}^K$ is defined by

$$\boldsymbol{z} = \boldsymbol{x}\boldsymbol{y} \iff z_k = x_k y_k, \forall k.$$

If $\mathcal{X}$ and $\mathcal{Y}$ are subsets of a vector space $\mathcal{V}$, the sum of $\mathcal{X}$ and $\mathcal{Y}$ is defined as follows:

$$\mathcal{X} + \mathcal{Y} = \{\boldsymbol{x} + \boldsymbol{y} : (\boldsymbol{x}, \boldsymbol{y}) \in \mathcal{X} \times \mathcal{Y})\}.$$

Moreover, if $\mathcal{V}$ is a real vector space and $\alpha \in \mathbb{R}$, then

$$\alpha\mathcal{X} = \{\alpha\boldsymbol{x} : \boldsymbol{x} \in \mathcal{X}\}.$$

A.1.3 Functions

Let $f : \mathcal{X} \to \mathcal{Y}$ be a function. The image of a set $\mathcal{A} \subseteq \mathcal{X}$ under f is given by

$$f(\mathcal{A}) = \{f(x) : x \in \mathcal{A}\}.$$

The preimage of a set $\mathcal{C} \subseteq \mathcal{Y}$ under f is

$$f^{-1}(\mathcal{C}) = \{x \in \mathcal{X} : f(x) \in \mathcal{C}\}.$$

If f is a bijection, its inverse function is denoted by f^{-1}. Accordingly, for $c \in \mathcal{Y}$, $f^{-1}(c)$ is the value of the inverse function of f at c, while $f^{-1}(\{c\})$ is the preimage of $\{c\}$ under f.

Given a set $\mathcal{X}$, the identity function on $\mathcal{X}$ is the function $\mathrm{id} : \mathcal{X} \to \mathcal{X}$,

$$\mathrm{id}(x) = x.$$

The restriction of a function $f : \mathcal{X} \to \mathcal{Y}$ to a set $\mathcal{W} \subseteq \mathcal{X}$ is the function $f|_{\mathcal{W}} : \mathcal{W} \to \mathcal{Y}$,

$$f|_{\mathcal{W}}(x) = f(x).$$

Let $f : \mathcal{X} \to \mathcal{Y}$ and $g : \mathcal{W} \to \mathcal{Z}$ be two functions. Provided that $f(\mathcal{X}) \subseteq \mathcal{W}$, the composite function $g \circ f$ is the function $g \circ f : \mathcal{X} \to \mathcal{Z}$,

$$(g \circ f)(x) = g(f(x)).$$

The function (f, g) is the function defined by $(f, g) : \mathcal{X} \times \mathcal{W} \to \mathcal{Y} \times \mathcal{Z}$,

$$(f, g)(x, w) = (f(x), g(w)).$$

The operator ∇ yields the gradient field of a differentiable function $f : \mathbb{R}^K \to \mathbb{R}$, i.e., $\nabla f(\boldsymbol{x})$ is the gradient of f at $\boldsymbol{x}$. Likewise, applying ∇ to a function

$\boldsymbol{f}: \mathbb{R}^K \to \mathbb{R}^N$ with differentiable component functions f_n yields a matrix $\nabla \boldsymbol{f}(\boldsymbol{x})$, with the n-th column corresponding to the gradient of the n-th component function at $\boldsymbol{x}$. Equivalently, $\nabla \boldsymbol{f}(\boldsymbol{x})$ is the transpose of the Jacobian of $\boldsymbol{f}$ at $\boldsymbol{x}$. A function $\boldsymbol{f}$: $\mathbb{R}^K \to \mathbb{R}^N$ whose component functions f_n are r- times continuously differentiable is said to be a C^r function.

Given a function $f : \mathcal{X} \times \mathcal{Y} \to \mathcal{Z}, (x, y) \mapsto f(x, y)$, a function $f(\cdot, y)$ is defined by

$$f(\cdot, y) : \mathcal{X} \to \mathcal{Z}, x \mapsto f(x, y), \tag{A.1}$$

where it is assumed that $y \in \mathcal{Y}$. Moreover, if $\mathcal{Z} \subseteq \mathbb{R}$, then $\min_{y \in \mathcal{Y}} f(\cdot, y)$ is the function defined by the point-wise minimum

$$\left(\min_{y \in \mathcal{Y}} f(\cdot, y)\right)(x) = \min_{y \in \mathcal{Y}} f(x, y), \tag{A.2}$$

where it is assumed that the minimum on the right-hand side of (A.2) is well-defined for all $x \in \mathcal{X}$.

For $\boldsymbol{x} \in \mathbb{R}^K$, $(\boldsymbol{x})^+$ is the vector in $\mathbb{R}^K_+$ obtained by setting the negative entries in $\boldsymbol{x}$ to zero.

A.1.4 Order Relations, Monotonicity, and Pareto Optimality

The following definitions of order relations between vectors $\boldsymbol{x}, \boldsymbol{y} \in \mathbb{R}^N$, with $N > 1$, are used:

$$\begin{aligned}
\boldsymbol{x} \geq \boldsymbol{y} &\iff \forall k : x_k \geq y_k, \\
\boldsymbol{x} > \boldsymbol{y} &\iff \boldsymbol{x} \geq \boldsymbol{y}, \exists k : x_k > y_k, \\
\boldsymbol{x} \gg \boldsymbol{y} &\iff \forall k : x_k > y_k.
\end{aligned}$$

Order relations $\leq, <, \ll$ are defined in the same manner.

A function $\boldsymbol{f} : \mathcal{X} \to \mathbb{R}^N$, with $\mathcal{X} \subseteq \mathbb{R}^M$, is said to be *increasing* if

$$\boldsymbol{x} \in \mathcal{X} \text{ and } \boldsymbol{x}' \geq \boldsymbol{x} \implies \boldsymbol{x}' \in \mathcal{X} \text{ and } \boldsymbol{f}(\boldsymbol{x}') \geq \boldsymbol{f}(\boldsymbol{x}).$$

Likewise, $\boldsymbol{f}$ is said to be *strictly increasing* if it is increasing and

$$\boldsymbol{x}' > \boldsymbol{x} \implies \boldsymbol{f}(\boldsymbol{x}') > \boldsymbol{f}(\boldsymbol{x}).$$

Finally, $\boldsymbol{f}$ is said to be *strongly increasing* if it is increasing and

$$\boldsymbol{x}' \gg \boldsymbol{x} \implies \boldsymbol{f}(\boldsymbol{x}') \gg \boldsymbol{f}(\boldsymbol{x}).$$

For a set $\mathcal{S} \subset \mathbb{R}^K$, the set of Pareto optimal points of $\mathcal{S}$ is defined by

$$\begin{aligned}\text{par}(\mathcal{S}) &= \left\{\boldsymbol{x} \in \mathcal{S} : (\{\boldsymbol{x}\} + \mathbb{R}^K_+) \cap \mathcal{S} = \{\boldsymbol{x}\}\right\} \\ &= \left\{\boldsymbol{x} \in \mathcal{S} : \nexists \boldsymbol{x}' \in \mathcal{S} : \boldsymbol{x}' > \boldsymbol{x}\right\}.\end{aligned}$$

Likewise, the set of weakly Pareto optimal points of $\mathcal{S}$ is given by

$$\begin{aligned}\text{wpar}(\mathcal{S}) &= \left\{\boldsymbol{x} \in \mathcal{S} : (\{\boldsymbol{x}\} + \mathbb{R}^K_{++}) \cap \mathcal{S} = \emptyset\right\} \\ &= \left\{\boldsymbol{x} \in \mathcal{S} : \nexists \boldsymbol{x}' \in \mathcal{S} : \boldsymbol{x}' \gg \boldsymbol{x}\right\}.\end{aligned}$$

A.2 Elementary Topology

This appendix primarily summarizes basic topological definitions and results that are used throughout this work. Accordingly, in most cases proofs are omitted. Proofs and further details can be found in any textbook on elementary topology, such as [85]. Familiarity with the following basic concepts is assumed: topological space, topological subspace, metric space, and natural topology in a metric space.

The minimal topological structure considered in this work is that of a metric space. Accordingly, the following definitions and propositions are formulated using metric spaces.

Definition A.2.1 (Continuous Function). Let $\mathcal{X}$ and $\mathcal{Y}$ be two metric spaces. A function $f : \mathcal{X} \to \mathcal{Y}$ is continuous if and only if the preimage of any open subset of $\mathcal{Y}$ is an open subset of $\mathcal{X}$.

Definition A.2.2 (Homeomorphism). An invertible function f is a *homeomorphism* if both f and f^{-1} are continuous.

Proposition A.2.1. *Suppose $f : \mathcal{X} \to \mathcal{Y}$ is a continuous function between two metric spaces $\mathcal{X}$ and $\mathcal{Y}$, and $\mathcal{Z} \subset \mathcal{Y}$ is closed. Then the preimage $f^{-1}(\mathcal{Z})$ is closed.*

Proposition A.2.2. *Let $f : \mathcal{X} \to \mathcal{Y}$ be a continuous function between two metric spaces $\mathcal{X}$ and $\mathcal{Y}$. If $\mathcal{X}$ is compact, then $f(\mathcal{X})$ is compact.*

Proposition A.2.3. *Suppose $\mathcal{X}$ is a metric space, $\mathcal{F} \subset \mathcal{X}$ is compact, and $\mathcal{G} \subset \mathcal{X}$ is closed. Then $\mathcal{F} \cap \mathcal{G}$ is compact.*

Theorem A.2.4 (Weierstrass). *Let $\mathcal{X}$ be a nonempty compact set, and let $f : \mathcal{X} \to \mathbb{R}$ be a continuous function. Then there exist $\bar{x}, \underline{x} \in \mathcal{X}$ such that*

$$f(\bar{x}) = \sup\{f(x) : x \in \mathcal{X}\}, \tag{A.3}$$

$$f(\underline{x}) = \inf\{f(x) : x \in \mathcal{X}\}. \tag{A.4}$$

Proposition A.2.5. *Let $\mathcal{X}$ and $\mathcal{Y}$ be metric spaces, with $\mathcal{X}$ nonempty and compact, and let $f : \mathcal{X} \times \mathcal{Y} \to \mathbb{R}$ be continuous. Then the function $g : \mathcal{Y} \to \mathbb{R}$, with*

$$g(\boldsymbol{y}) = \min_{\boldsymbol{x} \in \mathcal{X}} f(\boldsymbol{x}, \boldsymbol{y})$$

is well-defined and continuous.

Proof. The function f is continuous, thus the function $f(\cdot, \boldsymbol{y})$ is continuous, for any $\boldsymbol{y} \in \mathcal{Y}$. Moreover, $\mathcal{X}$ is compact. Consequently, the minimum $g(\boldsymbol{y})$ is attained on $\mathcal{X}$, thus g is well-defined.

Likewise, the function $f(\boldsymbol{x}, \cdot)$ is continuous for any $\boldsymbol{x} \in \mathcal{X}$. The function g corresponds to the point-wise minimum over the set of continuous functions $\{f(\boldsymbol{x}, \cdot)\}_{\boldsymbol{x} \in \mathcal{X}}$. Thus, g is continuous. □

Corollary A.2.6. *Let $\mathcal{Y}$ be a metric space, and let $\boldsymbol{h} : \mathcal{Y} \to \mathbb{R}^K$ be continuous. Then the function $g : \mathcal{Y} \to \mathbb{R}$, with*

$$g(\boldsymbol{y}) = \min_{k \in \{1,\ldots,K\}} h_k(\boldsymbol{y})$$

is continuous.

Proof. The result follows from Proposition A.2.5 by letting

$$\mathcal{X} = \{1, \ldots, K\},$$
$$f(x, \boldsymbol{y}) = h_x(\boldsymbol{y}),$$

and observing that $\mathcal{X}$ is compact and $f : \mathcal{X} \times \mathcal{Y} \to \mathbb{R}$ is continuous. □

Theorem A.2.7 (Heine-Borel). *Let $\mathcal{X} \subset \mathbb{R}^N$. Then the following statements are equivalent:*

(1) $\mathcal{X}$ is closed and bounded.
(2) $\mathcal{X}$ is compact.

A.3 Pareto Sets

Proposition A.3.1. *Let $\mathcal{X}$ denote a nonempty subset of $\mathbb{R}^K$. Then*

$$\operatorname{wpar}(\mathcal{X}) \subseteq \operatorname{bd}(\mathcal{X}). \tag{A.5}$$

Proof. Let $\boldsymbol{x} \in \operatorname{wpar}(\mathcal{X})$. Assume $\boldsymbol{x} \notin \operatorname{bd}(\mathcal{X})$. Then there exists $\eta > 0$ such that

$$\mathcal{B}_\eta(\boldsymbol{x}) \subset \mathcal{X},$$

which contradicts $\boldsymbol{x} \in \operatorname{wpar}(\mathcal{X})$. □

Proposition A.3.2. *Let $\mathcal{X}$ denote a nonempty compact subset of $\mathbb{R}_+^K$. Then*

$$\mathrm{par}\,(\mathcal{X}) \neq \emptyset. \tag{A.6}$$

Proof. Define

$$\bar{\mathcal{X}} = \operatorname*{argmax}_{\boldsymbol{x} \in \mathcal{X}} \|\boldsymbol{x}\|_2 .$$

The set $\mathcal{X}$ is compact and nonempty, and $\|\cdot\|_2$ is continuous, thus $\bar{\mathcal{X}}$ is well-defined and non-empty. Let $\boldsymbol{x}' \in \bar{\mathcal{X}}$. Assume $\boldsymbol{x}' \notin \mathrm{par}\,(\mathcal{X})$. Then there exists $\boldsymbol{v} \in \mathbb{R}_+^K \setminus \{\boldsymbol{0}\}$ such that $\boldsymbol{x}' + \boldsymbol{v} \in \mathcal{X}$. But

$$\left\|\boldsymbol{x}' + \boldsymbol{v}\right\|_2 = \sqrt{\boldsymbol{x}^{\mathrm{T}}\boldsymbol{x} + 2\boldsymbol{x}^{\mathrm{T}}\boldsymbol{v} + \boldsymbol{v}^{\mathrm{T}}\boldsymbol{v}} > \left\|\boldsymbol{x}'\right\|_2 ,$$

where the inequality follows from the fact that $\boldsymbol{x}, \boldsymbol{v} \in \mathbb{R}_+^K$ and $\boldsymbol{v} \neq \boldsymbol{0}$. This contradicts $\boldsymbol{x}' \in \bar{\mathcal{X}}$, thus

$$\mathrm{par}\,(\mathcal{X}) \supseteq \bar{\mathcal{X}} \supset \emptyset.$$

□

Corollary A.3.3. *Let $\mathcal{X}$ denote a nonempty compact subset of $\mathbb{R}_+^K$. Then*

$$\mathrm{wpar}\,(\mathcal{X}) \neq \emptyset. \tag{A.7}$$

Proposition A.3.4. *Let $\mathcal{X}$ denote a compact subset of $\mathbb{R}^K$. Then* $\mathrm{wpar}\,(\mathcal{X})$ *is compact.*

Proof. Let $\{\boldsymbol{x}^{(n)}\}$ denote a sequence with $\boldsymbol{x}^{(n)} \in \mathrm{wpar}\,(\mathcal{X})$ and

$$\lim_{n \to \infty} \boldsymbol{x}^{(n)} = \boldsymbol{x}.$$

By definition, $\mathrm{wpar}\,(\mathcal{X}) \subseteq \mathcal{X}$, thus $\boldsymbol{x}^{(n)} \in \mathcal{X}$. The set $\mathcal{X}$ is closed, hence $\boldsymbol{x} \in \mathcal{X}$. Assume that $\boldsymbol{x} \notin \mathrm{wpar}\,(\mathcal{X})$. Then there exists $\boldsymbol{x}' \in \mathcal{X}$ such that

$$\boldsymbol{x}' \gg \boldsymbol{x}. \tag{A.8}$$

Let

$$\delta = \min_k x_k' - x_k .$$

Equation (A.8) implies that $\delta > 0$. It follows that

$$\boldsymbol{x}' \gg \boldsymbol{y}, \forall \boldsymbol{y} \in \mathcal{B}_\delta(\boldsymbol{x}).$$

As a result, $\mathcal{B}_\delta(\boldsymbol{x})$ does not contain any weak Pareto optimal points, which contradicts the fact that $\boldsymbol{x}$ is a limit point of a sequence of weak Pareto optimal points. Thus, $\boldsymbol{x} \in \mathrm{wpar}\,(\mathcal{X})$, implying that $\mathrm{wpar}\,(\mathcal{X})$ is closed. As a subset of a bounded set, $\mathrm{wpar}\,(\mathcal{X})$ is bounded. Theorem A.2.7 yields the result. □

Proposition A.3.5. *Let $\mathcal{X} \subset \mathbb{R}_+^K$ denote a compact set and let $\phi : \mathbb{R} \to \mathbb{R}^K$ denote a continuous function. Moreover, assume that there exist $\underline{\gamma}$ and $\overline{\gamma}$ such that $\phi(\underline{\gamma}) \in \mathcal{X}$ and ϕ restricted to $\{\overline{\gamma}\} + \mathbb{R}_+$ is strictly increasing. Then the set*

$$\mathcal{G} = \{\gamma \in \mathbb{R} : \phi(\gamma) \in \mathcal{X}\}$$

is closed, nonempty, and bounded from above. Furthermore,

$$\phi(\gamma^*) \in \mathrm{bd}\,(\mathcal{X}),$$

where

$$\gamma^* = \max\{\gamma \in \mathbb{R} : \phi(\gamma) \in \mathcal{X}\}.$$

Proof. The function ϕ is continuous and the set $\mathcal{X}$ is closed, thus the reverse image $\mathcal{G} = \phi^{-1}(\mathcal{X})$ is closed. Obviously, $\left\{\underline{\gamma}\right\} \subseteq \mathcal{G}$. From ϕ restricted to $\overline{\gamma} + \mathbb{R}_+$ being strictly increasing it follows that

$$\gamma' > \gamma \implies \left\|\phi(\gamma')\right\|_2 > \|\phi(\gamma)\|_2, \forall \gamma \geq \overline{\gamma}. \tag{A.9}$$

The set $\mathcal{X}$ is bounded, thus (A.9) implies that $\mathcal{G}$ is bounded from above. Assume that $\phi(\gamma^*) \notin \mathrm{bd}\,(\mathcal{X})$. This implies there exists $\eta > 0$ such that

$$\mathcal{B}_\eta(\phi(\gamma^*)) \in \mathcal{X},$$

which by continuity of ϕ imlies there exists $\epsilon > 0$ such that

$$\phi(\gamma^* + \epsilon) \in \mathcal{X},$$

which contradicts the definition of γ^*. □

Proposition A.3.6. *Let $\mathcal{X}$ denote a nonempty compact subset of $\mathbb{R}_+^K$. Let $\boldsymbol{x} \in \mathcal{X}$. Then there exists $\boldsymbol{x}^* \in \mathrm{par}\,(\mathcal{X})$ with*

$$\boldsymbol{x}^* \geq \boldsymbol{x}.$$

Proof. Let $\boldsymbol{x}^{(0)} = \boldsymbol{x}$. Define a sequence $\left\{\boldsymbol{x}^{(n)}\right\}_{n=0}^{\infty}$ as follows: If $\boldsymbol{x}^{(n)} \in \mathrm{par}\,(\mathcal{X})$, set

$$\boldsymbol{x}^{(n+1)} = \boldsymbol{x}^{(n)}. \tag{A.10}$$

If $\boldsymbol{x}^{(n)} \notin \mathrm{par}\,(\mathcal{X})$, there exists $\boldsymbol{x}' \in \mathcal{X}$ such that $\boldsymbol{x}' > \boldsymbol{x}^{(n)}$. Let

$$\boldsymbol{v} = \frac{\boldsymbol{x}' - \boldsymbol{x}^{(n)}}{\left\|\boldsymbol{x}' - \boldsymbol{x}^{(n)}\right\|_2},$$

and define

$$\gamma^{(n)} = \max_{\gamma \in \mathbb{R}} \gamma \quad \text{s.t.} \quad \boldsymbol{x}^{(n)} + \gamma \boldsymbol{v} \in \mathcal{X}.$$

Set

$$\boldsymbol{x}^{(n+1)} = \boldsymbol{x}^{(n)} + \gamma^{(n)} \boldsymbol{v}. \tag{A.11}$$

By Proposition A.3.5 $\gamma^{(n)}$ is well-defined and $\boldsymbol{x}^{(n+1)} \in \mathcal{X}$. Moreover, $\boldsymbol{x}' \in \mathcal{X}$ and $\boldsymbol{x}' > \boldsymbol{x}^{(n)}$ implies that

$$\gamma^{(n)} \geq \left\| \boldsymbol{x}' - \boldsymbol{x}^{(n)} \right\|_2 > 0.$$

From (A.10) and (A.11), it follows that

$$\boldsymbol{x}^{(n+1)} \geq \boldsymbol{x}^{(n)}. \tag{A.12}$$

This implies

$$\left\| \boldsymbol{x}^{(n+k)} - \boldsymbol{x}^{(n)} \right\|_2 \geq \left\| \boldsymbol{x}^{(n+1)} - \boldsymbol{x}^{(n)} \right\|_2, \forall k \geq 1. \tag{A.13}$$

The set $\mathcal{X}$ is compact, thus the sequence $\left\{ \boldsymbol{x}^{(n)} \right\}_{n=0}^{\infty}$ has a convergent subsequence with limit point $\boldsymbol{x}^* \in \mathcal{X}$. By (A.12),

$$\boldsymbol{x}^* \geq \boldsymbol{x}^{(0)} = \boldsymbol{x}.$$

Now, if $\boldsymbol{x}^{(n)} \notin \text{par}(\mathcal{X})$,

$$\left\| \boldsymbol{x}^{(n+1)} - \boldsymbol{x}^{(n)} \right\|_2 = \gamma^{(n)} > 0.$$

Thus,

$$\left\| \boldsymbol{x}^{(n+k)} - \boldsymbol{x}^{(n)} \right\|_2 \geq \gamma^{(n)}, \forall k \geq 1.$$

As a consequence, no point in $\mathcal{X} \setminus \text{par}(\mathcal{X})$ can be a limit point, implying that $\boldsymbol{x}^* \in \text{par}(\mathcal{X})$. □

A.4 Comprehensive Sets

Definition A.4.1 (Comprehensive Set). A subset $\mathcal{X}$ of $\mathbb{R}_+^K$ is comprehensive if

$$\boldsymbol{x} \in \mathcal{X}, \boldsymbol{x}' \in \mathbb{R}_+^K, \boldsymbol{x}' \leq \boldsymbol{x} \implies \boldsymbol{x}' \in \mathcal{X}.$$

Corollary A.4.1. *Let $\mathcal{X}$ be a nonempty comprehensive set. Then $\mathbf{0} \in \mathcal{X}$.*

Definition A.4.2 (Comprehensive Hull). The comprehensive hull of a set $\mathcal{X} \subseteq \mathbb{R}_+^K$ is the set

$$\text{comp}(\mathcal{X}) = \bigcup_{\boldsymbol{x} \in \mathcal{X}} \left\{ \boldsymbol{y} \in \mathbb{R}_+^K : \boldsymbol{y} \leq \boldsymbol{x} \right\}.$$

Corollary A.4.2.

$$\mathrm{comp}\,(\mathcal{X} \cup \mathcal{Y}) = \mathrm{comp}\,(\mathcal{X}) \cup \mathrm{comp}\,(\mathcal{Y})\,. \tag{A.14}$$

Proposition A.4.3. *Let $\mathcal{X}$ denote a comprehensive subset of $\mathbb{R}_+^K$ and let $\phi : \mathbb{R} \to \mathbb{R}_+^K$ denote a continuous increasing function. Moreover, assume that there exists a pair $(\gamma', \boldsymbol{s}) \in \mathbb{R} \times \mathcal{X}$ such that $\phi(\gamma') \ll \boldsymbol{s}$. Then there exists $\epsilon > 0$ such that*

$$\phi(\gamma' + \epsilon) \in \mathcal{X}.$$

Proof. Let

$$\mathcal{K}_+\,(\boldsymbol{x}) = \{\boldsymbol{x}\} + \mathbb{R}_+^K.$$

Due to the fact that ϕ is increasing,

$$\phi(\gamma' + \epsilon) \in \mathcal{K}_+\left(\phi(\gamma')\right), \forall \epsilon \geq 0. \tag{A.15}$$

By continuity of ϕ, for all $\eta > 0$ there exists $\epsilon > 0$ such that

$$\phi(\gamma' + \epsilon) \in \mathcal{B}_\eta(\phi(\gamma')). \tag{A.16}$$

Combining (A.15) and (A.16), it follows that for all $\eta > 0$ there exists $\epsilon > 0$ such that

$$\phi(\gamma' + \epsilon) \in \mathcal{B}_\eta(\phi(\gamma')) \cap \mathcal{K}_+\left(\phi(\gamma')\right). \tag{A.17}$$

From the comprehensiveness of $\mathcal{X}$, $\boldsymbol{s} \in \mathcal{X}$, $\phi(\gamma') \in \mathbb{R}_+^K$, and $\phi(\gamma') \leq \boldsymbol{s}$, it follows that

$$\left\{\boldsymbol{r} \in \mathbb{R}_+^K : \phi(\gamma') \leq \boldsymbol{r} \leq \boldsymbol{s}\right\} \subseteq \mathcal{X}.$$

Moreover, $\phi(\gamma') \ll \boldsymbol{s}$ implies that there exists $\eta > 0$ such that

$$\mathcal{B}_\eta(\phi(\gamma')) \cap \mathcal{K}_+\left(\phi(\gamma')\right) \subseteq \mathcal{X}. \tag{A.18}$$

Combined with (A.17), (A.18) yields the result. □

Note that the existence of a pair $(\gamma', \boldsymbol{s}) \in \mathbb{R} \times \mathcal{X}$ with $\phi(\gamma') \ll \boldsymbol{s}$ does not imply $\mathcal{B}_\eta(\phi(\gamma')) \subset \mathcal{X}$. As an example, take $\phi(\gamma') = \mathbf{0}$ and $\boldsymbol{s} = \mathbf{1}$. Then $\mathcal{B}_1(\phi(\gamma')) \cap \mathcal{K}_+\,(\phi(\gamma')) \subset \mathcal{X}$, while $\mathcal{B}_1(\phi(\gamma')) \not\subset \mathcal{X}$. In other words, the property that ϕ is increasing is needed to establish the result.

Proposition A.4.4. *Let $\mathcal{X} \subset \mathbb{R}_+^K$ denote a compact comprehensive set and let $\phi : \mathbb{R} \to \mathbb{R}^K$ denote a continuous function. Moreover, assume that there exist $\underline{\gamma}$ and $\overline{\gamma}$ such that $\phi(\underline{\gamma}) \in \mathcal{X}$ and ϕ restricted to $\overline{\gamma} + \mathbb{R}_+$ is strictly increasing. Then*

$$\phi(\gamma^*) \in \mathrm{wpar}\,(\mathcal{X})\,,$$

where

$$\gamma^* = \max\{\gamma \in \mathbb{R} : \phi(\gamma) \in \mathcal{X}\}.$$

Proof. From Proposition A.3.5 it follows that γ^* is well-defined and $\phi(\gamma^*) \in \mathcal{X}$.

Assume that $\phi(\gamma^*) \notin \mathrm{wpar}(\mathcal{X})$. By definition of $\mathrm{wpar}(\mathcal{X})$, this implies that there exists $\boldsymbol{s} \in \mathcal{X}$ with $\boldsymbol{s} \gg \phi(\gamma^*)$. By Proposition A.4.3, this implies that there exists $\epsilon > 0$ such that $\phi(\gamma^* + \epsilon) \in \mathcal{X}$, which contradicts the definition of γ^*. □

Proposition A.4.5. *Let $\mathcal{X}$ be a compact comprehensive set. Then*

$$\mathcal{X} = \mathrm{comp}(\mathrm{wpar}(\mathcal{X})). \tag{A.19}$$

Proof. If $\mathcal{X}$ is empty, the equation trivially holds. Assume $\mathcal{X}$ is nonempty. Then $\mathcal{X}$ is a nonempty compact subset of $\mathbb{R}_+^K$, thus, by Corollary A.3.3, $\mathrm{wpar}(\mathcal{X})$ is nonempty.

The relation $\mathrm{comp}(\mathrm{wpar}(\mathcal{X})) \subseteq \mathcal{X}$ follows from $\mathrm{wpar}(\mathcal{X}) \subseteq \mathcal{X}$, thus $\mathrm{comp}(\mathrm{wpar}(\mathcal{X})) \subseteq \mathrm{comp}(\mathcal{X}) = \mathcal{X}$.

To show $\mathcal{X} \subseteq \mathrm{comp}(\mathrm{wpar}(\mathcal{X}))$, let $\boldsymbol{x} \in \mathcal{X}$. If $\boldsymbol{x} = \boldsymbol{0}$, then $\boldsymbol{x} \in \mathrm{comp}(\mathrm{wpar}(\mathcal{X}))$, as $\mathrm{comp}(\mathrm{wpar}(\mathcal{X}))$ is a nonempty comprehensive set. If $\boldsymbol{x} \neq \boldsymbol{0}$, then by Proposition A.4.4 $\gamma^*\boldsymbol{x} \in \mathrm{wpar}(\mathcal{X})$, with γ^* as defined in Proposition A.4.4. By definition of γ^*, $\boldsymbol{x} \leq \gamma^*\boldsymbol{x}$, thus $\boldsymbol{x} \in \mathrm{comp}(\mathrm{wpar}(\mathcal{X}))$. □

Proposition A.4.6. *Let $\mathcal{X} \subseteq \mathbb{R}_+^K$ be comprehensive. Then*

$$\mathrm{par}(\mathcal{X} \cap \mathcal{Z}_{\mathcal{D}}) \subseteq \mathrm{wpar}(\mathcal{X}),$$

where $\mathcal{D} \subset \{1, \ldots, K\}$ and

$$\mathcal{Z}_{\mathcal{D}} = \left\{\boldsymbol{x} \in \mathbb{R}_+^K : x_k = 0, k \in \mathcal{D}\right\} \cap \left\{\boldsymbol{x} \in \mathbb{R}_+^K : x_k > 0, k \in \{1, \ldots, K\} \setminus \mathcal{D}\right\}.$$

Proof. Let $\boldsymbol{x} \in \mathrm{par}(\mathcal{X} \cap \mathcal{Z}_{\mathcal{D}})$. Assume that $\boldsymbol{x} \notin \mathrm{wpar}(\mathcal{X})$. This implies there exists $\boldsymbol{y} \in \mathcal{X}$ such that $\boldsymbol{y} \gg \boldsymbol{x}$. Let

$$z_k = \begin{cases} 0, k \in \mathcal{D}, \\ y_k, k \notin \mathcal{D}. \end{cases}$$

By comprehensiveness of $\mathcal{X}$, $\boldsymbol{z} \in \mathcal{X}$. But $\boldsymbol{z} \in \mathcal{Z}_{\mathcal{D}}$ and $\boldsymbol{z} > \boldsymbol{y}$, which contradicts $\boldsymbol{x} \in \mathrm{par}(\mathcal{X} \cap \mathcal{Z}_{\mathcal{D}})$. □

A.5 Convex Analysis

Let $\mathcal{V}$ denote a real or complex vector space.

Definition A.5.1. A set $\mathcal{C} \subseteq \mathcal{V}$ is convex if

$$\alpha\boldsymbol{x} + (1-\alpha)\boldsymbol{y} \in \mathcal{C}, \ \forall(\boldsymbol{x}, \boldsymbol{y}, \alpha) \in \mathcal{C} \times \mathcal{C} \times (0, 1).$$

Definition A.5.2. Let $\mathcal{C} \subseteq \mathcal{V}$ be convex. A function $f : \mathcal{C} \to \mathbb{R}^M$, with $M \geq 1$, is concave if

$$f(\alpha \boldsymbol{x} + (1-\alpha)\boldsymbol{y}) \geq \alpha \boldsymbol{f}(\boldsymbol{x}) + (1-\alpha)\boldsymbol{f}(\boldsymbol{y}), \ \forall (\boldsymbol{x}, \boldsymbol{y}, \alpha) \in \mathcal{C} \times \mathcal{C} \times (0,1).$$

Proposition A.5.1. *Let $\mathcal{C} \subseteq \mathbb{R}^N$ be convex. A concave function $\boldsymbol{f} : \mathcal{C} \to \mathbb{R}^M$ is continuous.*

Proof. By concavity of $\boldsymbol{f} = (f_1, \ldots, f_M)$, the component functions $f_m : \mathbb{R}^N \to \mathbb{R}$ are concave. A concave function $f_m : \mathcal{C} \to \mathbb{R}$ is continuous [28], thus $\boldsymbol{f}$ is continuous. □

Proposition A.5.2. *Let $\boldsymbol{g} : \mathcal{C} \to \mathbb{R}^M$ and $\boldsymbol{h} : \mathcal{D} \to \mathbb{R}^N$ be concave functions. Moreover, let $\boldsymbol{h}$ be increasing and $\boldsymbol{g}(\mathcal{C}) \subseteq \mathcal{D}$. Then the composite function $\boldsymbol{h} \circ \boldsymbol{g}$ is concave.*

Proof. Let $\boldsymbol{x}, \boldsymbol{y} \in \mathcal{C}$ and $\alpha \in (0,1)$. Due to the concavity of $\boldsymbol{g}$,

$$\boldsymbol{g}(\alpha \boldsymbol{x} + (1-\alpha)\boldsymbol{y}) \geq \alpha \boldsymbol{g}(\boldsymbol{x}) + (1-\alpha)\boldsymbol{g}(\boldsymbol{y}). \tag{A.20}$$

Hence,

$$\begin{aligned} \boldsymbol{h}(\boldsymbol{g}(\alpha \boldsymbol{x} + (1-\alpha)\boldsymbol{y})) &\geq \boldsymbol{h}(\alpha \boldsymbol{g}(\boldsymbol{x}) + (1-\alpha)\boldsymbol{g}(\boldsymbol{y})) \\ &\geq \alpha \boldsymbol{h}(\boldsymbol{g}(\boldsymbol{x})) + (1-\alpha)\boldsymbol{h}(\boldsymbol{g}(\boldsymbol{y})), \end{aligned}$$

where the first inequality follows from (A.20) and the fact that $\boldsymbol{h}$ is increasing, while the second inequality follows from the concavity of $\boldsymbol{h}$. □

Definition A.5.3 (Convex Hull). Let $\mathcal{X} \subseteq \mathcal{V}$. The convex hull of $\mathcal{X}$ is the smallest convex set containing $\mathcal{X}$,

$$\mathrm{conv}(\mathcal{X}) = \cap \{\mathcal{S} \subseteq \mathcal{V} : \mathcal{S} \text{ is convex and } \mathcal{X} \subseteq \mathcal{S}\}.$$

Theorem A.5.3. *Let $\mathcal{X} \subset \mathbb{R}^K$ be compact. Then* conv $(\mathcal{X})$ *is compact.*

Proof. [22, Theorem 1.4.3]. □

Proposition A.5.4. *Let $\mathcal{X} \subset \mathbb{R}^K$ be comprehensive. Then $\mathcal{C} = \mathrm{conv}(\mathcal{X})$ is comprehensive.*

Proof. By Carathéodory's Theorem [22], each point in $\mathcal{C}$ can be written as a convex combination of at most $K+1$ points in $\mathcal{X}$. Accordingly, for any $\boldsymbol{s} \in \mathcal{C}$ there exist coefficients $b_k \geq 0$ and points $\boldsymbol{s}^k \in \mathcal{X}$ such that $\sum_{k=1}^{K+1} b_k = 1$ and

$$\boldsymbol{s} = \sum_{k=1}^{K+1} b_k \boldsymbol{s}^k.$$

Let $s \in \mathcal{C}$ and $\mathbf{0} \leq s' \leq s$. There exists $a \in \mathbb{R}_+^K$ with $a \leq \mathbf{1}$ such that $s' = as$. Hence,

$$s' = \sum_{k=1}^{K+1} b_k a s^k.$$

By comprehensiveness of $\mathcal{X}$, $as^k \in \mathcal{X}$, thus $s' \in \mathcal{C}$. □

Proposition A.5.5. *Let $x \in \mathrm{bd}(\mathcal{C})$, where $\mathcal{C}$ is a nonempty compact convex subset of $\mathbb{R}^K$. Then there exists a hyperplane supporting $\mathcal{C}$ at x, i.e.,*

$$\exists \boldsymbol{\lambda} \in \mathbb{R}^K \setminus \{\mathbf{0}\} : \boldsymbol{\lambda}^{\mathrm{T}} x = \max_{y \in \mathcal{C}} \boldsymbol{\lambda}^{\mathrm{T}} y.$$

Proof. [22, Chap. III, Lemma 4.2.1]. □

Proposition A.5.6. *Let $x \in \mathrm{wpar}(\mathcal{C})$, where $\mathcal{C}$ is a nonempty compact convex subset of $\mathbb{R}_+^K$. Then*

$$\exists \boldsymbol{\lambda} \in \mathbb{R}_+^K \setminus \{\mathbf{0}\} : \boldsymbol{\lambda}^{\mathrm{T}} x = \max_{y \in \mathcal{C}} \boldsymbol{\lambda}^{\mathrm{T}} y.$$

Proof. Define a set $\mathcal{A} = \mathrm{comp}(\mathcal{C} + \{\mathbf{1}\})$. The set $\mathcal{A}$ is convex and comprehensive. Let $x \in \mathrm{wpar}(\mathcal{C})$. Define $v = x + \mathbf{1}$. Clearly, $v \in \mathbb{R}_{++}^K$. Moreover, $v \in \mathrm{wpar}(\mathcal{A})$, thus $v \in \mathrm{bd}(\mathcal{A})$. By Proposition A.5.5 there exists $\boldsymbol{\lambda} \in \mathbb{R}^K \setminus \{\mathbf{0}\}$ such that

$$\boldsymbol{\lambda}^{\mathrm{T}} v = \max_{y \in \mathcal{A}} \boldsymbol{\lambda}^{\mathrm{T}} y. \tag{A.21}$$

Define $z \in \mathbb{R}^K$ as follows:

$$z_k = \begin{cases} v_k, \lambda_k \geq 0, \\ 0, \lambda_k < 0. \end{cases} \tag{A.22}$$

Observe that $\mathbf{0} \leq z \leq v$, $v \in \mathcal{A}$, and $\mathcal{A}$ is comprehensive, hence $z \in \mathcal{A}$. Assume that there exists k such that $\lambda_k < 0$. Then $\boldsymbol{\lambda}^{\mathrm{T}} v < \boldsymbol{\lambda}^{\mathrm{T}} z$, which contradicts (A.21). Thus, $\boldsymbol{\lambda} \in \mathbb{R}_+^K \setminus \{\mathbf{0}\}$. Moreover,

$$\begin{aligned} \boldsymbol{\lambda}^{\mathrm{T}} x + \boldsymbol{\lambda}^{\mathrm{T}} \mathbf{1} &= \boldsymbol{\lambda}^{\mathrm{T}} v \\ &= \max_{y \in \mathcal{A}} \boldsymbol{\lambda}^{\mathrm{T}} y \\ &\geq \max_{y \in (\mathcal{C} + \{\mathbf{1}\})} \boldsymbol{\lambda}^{\mathrm{T}} y \\ &= \max_{y \in \mathcal{C}} \boldsymbol{\lambda}^{\mathrm{T}} y + \boldsymbol{\lambda}^{\mathrm{T}} \mathbf{1}. \end{aligned}$$

Subtracting $\boldsymbol{\lambda}^{\mathrm{T}} \mathbf{1}$ on both sides and noting that $x \in \mathcal{C}$ yields the result. □

Proposition A.5.7. *Let $\mathcal{X}$ denote a compact subset of $\mathbb{R}^K$. Any $\boldsymbol{x} \in \operatorname{bd}(\operatorname{conv}(\mathcal{X}))$ can be represented as a convex combination of at most K elements of $\mathcal{X}$.*

Proof. [22, Chap. III, Proposition 4.2.3]. □

Proposition A.5.8. *Let $\mathcal{C}$ denote a convex compact set. Let $\boldsymbol{s}$ be a convex combination of W elements of $\mathcal{C}$, i.e.,*

$$\boldsymbol{s} = \sum_{w=1}^{W} \alpha_w \boldsymbol{s}^{(w)}, \tag{A.23}$$

with $\boldsymbol{s}^{(w)} \in \mathcal{C}$, $\alpha_w > 0$, and $\sum_{w=1}^{W} \alpha_w = 1$. Moreover, assume that $\boldsymbol{s}$ maximizes $\boldsymbol{\lambda}^{\mathrm{T}}\boldsymbol{y}$ over $\mathcal{C}$, i.e.,

$$\boldsymbol{\lambda}^{\mathrm{T}}\boldsymbol{s} = \max_{\boldsymbol{y}\in\mathcal{C}} \boldsymbol{\lambda}^{\mathrm{T}}\boldsymbol{y}. \tag{A.24}$$

Then

$$\boldsymbol{\lambda}^{\mathrm{T}}\boldsymbol{s} = \boldsymbol{\lambda}^{\mathrm{T}}\boldsymbol{s}^{(w)}, \forall w = 1, \ldots, W.$$

Proof. First note that $\boldsymbol{\lambda}^{\mathrm{T}}\boldsymbol{s} < \boldsymbol{\lambda}^{\mathrm{T}}\boldsymbol{s}^{(w)}$ contradicts (A.24). Next, assume that there exist $q, r \in \{1, \ldots, W\}$ such that $\boldsymbol{\lambda}^{\mathrm{T}}\boldsymbol{s}^{(q)} < \boldsymbol{\lambda}^{\mathrm{T}}\boldsymbol{s}^{(r)}$. Let

$$\boldsymbol{s}' = \sum_{\substack{w=1\\ w\neq q}}^{W} \alpha_w \boldsymbol{s}^{(w)} + \alpha_q \boldsymbol{s}^{(r)}.$$

Then

$$\boldsymbol{\lambda}^{\mathrm{T}}\boldsymbol{s} < \boldsymbol{\lambda}^{\mathrm{T}}\boldsymbol{s}'.$$

This again contradicts (A.24), as, by convexity of $\mathcal{C}$, $\boldsymbol{s}' \in \mathcal{C}$. Thus,

$$\boldsymbol{\lambda}^{\mathrm{T}}\boldsymbol{s}^{(q)} = \boldsymbol{\lambda}^{\mathrm{T}}\boldsymbol{s}^{(r)}$$

and

$$\boldsymbol{\lambda}^{\mathrm{T}}\boldsymbol{s} = \sum_{w=1}^{W} \alpha_w \boldsymbol{\lambda}^{\mathrm{T}}\boldsymbol{s}^{(r)} = \boldsymbol{\lambda}^{\mathrm{T}}\boldsymbol{s}^{(r)} \sum_{w=1}^{W} \alpha_w = \boldsymbol{\lambda}^{\mathrm{T}}\boldsymbol{s}^{(r)}.$$

□

Corollary A.5.9. *Let $\mathcal{R}$ be a nonempty compact subset of $\mathbb{R}_+^K$. Let $\mathcal{C} = \operatorname{conv}(\mathcal{R})$ and $\boldsymbol{\lambda} \in \mathbb{R}_+^K$. Then*

$$\max_{\boldsymbol{r}\in\mathcal{R}} \boldsymbol{\lambda}^{\mathrm{T}}\boldsymbol{r} = \max_{\boldsymbol{r}\in\mathcal{C}} \boldsymbol{\lambda}^{\mathrm{T}}\boldsymbol{r},$$

$$\operatorname*{argmax}_{\boldsymbol{r}\in\mathcal{R}} \boldsymbol{\lambda}^{\mathrm{T}}\boldsymbol{r} \subseteq \operatorname*{argmax}_{\boldsymbol{r}\in\mathcal{C}} \boldsymbol{\lambda}^{\mathrm{T}}\boldsymbol{r}.$$

Proposition A.5.10. *Let* $\mathcal{R} \subset \mathbb{R}^K_+$ *be nonconvex, compact, and comprehensive. Let* $\mathcal{C} = \mathrm{conv}(\mathcal{R})$. *Then there exists a weakly Pareto optimal point of* $\mathcal{C}$ *that is not in* $\mathcal{R}$, *i.e.,*

$$\mathrm{wpar}(\mathcal{C}) \setminus (\mathrm{wpar}(\mathcal{C}) \cap \mathcal{R}) \neq \emptyset.$$

Proof. By nonconvexity of $\mathcal{R}$, $\mathcal{R} \subset \mathcal{C}$. The set $\mathcal{C}$ is compact and comprehensive, thus, by Proposition A.4.5, $\mathcal{C} = \mathrm{comp}(\mathrm{wpar}(\mathcal{C}))$. Using the equality

$$\mathrm{wpar}(\mathcal{C}) = (\mathrm{wpar}(\mathcal{C}) \cap \mathcal{R}) \cup (\mathrm{wpar}(\mathcal{C}) \setminus (\mathrm{wpar}(\mathcal{C}) \cap \mathcal{R})) \tag{A.25}$$

yields

$$\mathcal{C} = \mathrm{comp}(\mathrm{wpar}(\mathcal{C}) \cap \mathcal{R}) \cup \mathrm{comp}(\mathrm{wpar}(\mathcal{C}) \setminus (\mathrm{wpar}(\mathcal{C}) \cap \mathcal{R}))\,. \tag{A.26}$$

But $\mathrm{comp}(\mathrm{wpar}(\mathcal{C}) \cap \mathcal{R}) \subseteq \mathrm{comp}(\mathcal{R}) = \mathcal{R}$. As a result,

$$\mathcal{R} \subset \mathcal{C} \subseteq \mathcal{R} \cup \mathrm{comp}(\mathrm{wpar}(\mathcal{C}) \setminus (\mathrm{wpar}(\mathcal{C}) \cap \mathcal{R}))\,. \tag{A.27}$$

Thus, $\mathrm{wpar}(\mathcal{C}) \setminus (\mathrm{wpar}(\mathcal{C}) \cap \mathcal{R}) \neq \emptyset$. □

A.6 The Set of Extended Real Numbers

The set of extended real numbers is obtained by adding the elements "$-\infty$" and "∞" to the set of real numbers:

$$\tilde{\mathbb{R}} = \mathbb{R} \cup \{-\infty\} \cup \{\infty\}\,.$$

The following rules apply:

$$\begin{aligned}
-\infty \leq x \leq \infty,\ \forall x &\in \mathbb{R} \cup \{-\infty\} \cup \{\infty\}\,,\\
-1 \cdot (-\infty) &= \infty,\\
x + (-\infty) &= -\infty,\ \forall x \in \mathbb{R} \cup \{-\infty\}\,,\\
x + \infty &= \infty,\ \forall x \in \mathbb{R} \cup \{\infty\}\,,\\
x \cdot (-\infty) &= -\infty,\ \forall x \in \mathbb{R}_{++} \cup \{\infty\}\,,\\
x \cdot \infty &= \infty,\ \forall x \in \mathbb{R}_{++} \cup \{\infty\}\,.
\end{aligned}$$

Using the set of extended real numbers, it is possible to assign a value to the infimum and supremum of an empty set:

$$\inf \emptyset = \infty,$$
$$\sup \emptyset = -\infty.$$

A.7 Optimality Conditions

Definition A.7.1. Let $\mathcal{X}$ be a nonempty set, and $f : \mathcal{X} \to \mathbb{R}$ a function. Let $\boldsymbol{x}^* \in \mathcal{X}$. If

$$f(\boldsymbol{x}^*) \geq f(\boldsymbol{x}), \forall \boldsymbol{x} \in \mathcal{X}$$

then $f(\boldsymbol{x}^*)$ is the *global maximum* in $f(\mathcal{X})$, and $\boldsymbol{x}^*$ is a *global maximizer* of f in $\mathcal{X}$.

Definition A.7.2. Let $\mathcal{X}$ be a nonempty set, and $f : \mathcal{X} \to \mathbb{R}$ a function. Let $\boldsymbol{x}^* \in \mathcal{X}$. If

$$f(\boldsymbol{x}^*) + \epsilon \geq f(\boldsymbol{x}), \forall \boldsymbol{x} \in \mathcal{X}$$

then $f(\boldsymbol{x}^*)$ is a *global* ϵ*-maximum* in $f(\mathcal{X})$, and $\boldsymbol{x}^*$ is a *global* ϵ*-maximizer* of f in $\mathcal{X}$.

Definition A.7.3. Let $\mathcal{D}$ be a metric space, $f : \mathcal{D} \to \mathbb{R}$, and $\mathcal{X}$ a nonempty subset of $\mathcal{D}$. Let $\boldsymbol{x}^* \in \mathcal{X}$. If there exists $\epsilon > 0$ such that

$$f(\boldsymbol{x}^*) \geq f(\boldsymbol{x}), \forall \boldsymbol{x} \in \mathcal{B}_\epsilon(\boldsymbol{x}^*) \cap \mathcal{X}$$

then $f(\boldsymbol{x}^*)$ is a *local maximum* in $f(\mathcal{X})$ and $\boldsymbol{x}^*$ a *local maximizer* of f in $\mathcal{X}$.

Definition A.7.4. Let $\mathcal{X}$ be a subset of $\mathbb{R}^K$. A vector $\boldsymbol{v} \in \mathbb{R}^K$ is a *tangent direction* of $\mathcal{X}$ at $\boldsymbol{x}$ if there exists a sequence $\{\boldsymbol{x}^{(n)}\}$, with $\boldsymbol{x}^{(n)} \in \mathcal{X}$, and a sequence $\{t^{(n)}\}$, $t^{(n)} \in \mathbb{R}_{++}$, such that

$$\lim_{n\to\infty} \boldsymbol{x}^{(n)} = \boldsymbol{x}, \tag{A.28}$$

$$\lim_{n\to\infty} t^{(n)} = 0, \tag{A.29}$$

$$\lim_{n\to\infty} \frac{\boldsymbol{x}^{(n)} - \boldsymbol{x}}{t^{(n)}} = \boldsymbol{v}. \tag{A.30}$$

The set of all tangent directions of $\mathcal{X}$ at $\boldsymbol{x}$ is the *tangent cone* of $\mathcal{X}$ at $\boldsymbol{x}$, denoted by $\mathcal{T}_\mathcal{X}(\boldsymbol{x})$.

Corollary A.7.1. *Let* $\mathcal{S} = \mathcal{X} \cap \mathcal{Y}$. *Then*

$$\mathcal{T}_\mathcal{S}(\boldsymbol{x}) \subseteq \mathcal{T}_\mathcal{X}(\boldsymbol{x}) \cap \mathcal{T}_\mathcal{Y}(\boldsymbol{x}).$$

Proposition A.7.2. *Let $\mathcal{X} \subset \mathbb{R}^K$. Then the vectors in the tangent cone of the weak Pareto set of $\mathcal{X}$ at $\boldsymbol{x}$ are neither strictly positive nor strictly negative:*

$$\mathcal{T}_{\text{wpar}(\mathcal{X})}(\boldsymbol{x}) \subset \mathbb{R}^K \setminus (\mathbb{R}^K_{++} \cup -\mathbb{R}^K_{++}), \forall \boldsymbol{x} \in \mathbb{R}^K.$$

Proof. Assume that $\mathcal{T}_{\text{wpar}(\mathcal{X})}(\boldsymbol{x})$ is nonempty, as otherwise the result is trivial. Let $\boldsymbol{v} \in \mathbb{R}^K$ be a tangent direction of wpar $(\mathcal{X})$ at $\boldsymbol{x}$. Then there exists a sequence $\{\boldsymbol{x}^{(n)}\}$, with $\boldsymbol{x}^{(n)} \in$ wpar $(\mathcal{X})$, and a sequence $\{t^{(n)}\}$, $t^{(n)} \in \mathbb{R}_{++}$, such that

$$\lim_{n\to\infty} \boldsymbol{x}^{(n)} = \boldsymbol{x},$$

$$\lim_{n\to\infty} t^{(n)} = 0,$$

$$\lim_{n\to\infty} \frac{\boldsymbol{x}^{(n)} - \boldsymbol{x}}{t^{(n)}} = \boldsymbol{v}.$$

The assumption $\boldsymbol{v} \in \mathbb{R}^K_{++}$ implies that there exists n' such that $\boldsymbol{x}^{(n)} \gg \boldsymbol{x}, \forall n \geq n'$. This contradicts $\boldsymbol{x} \in$ wpar $(\mathcal{X})$. Likewise, the assumption $\boldsymbol{v} \in -\mathbb{R}^K_{++}$ implies that there exists n' such that $\boldsymbol{x}^{(n)} \ll \boldsymbol{x}, \forall n \geq n'$. This contradicts $\boldsymbol{x}^{(n)} \in$ wpar $(\mathcal{X})$. □

Theorem A.7.3. *Let $\mathcal{X}$ be a nonempty subset of $\mathbb{R}^K$, $f : \mathcal{D} \to \mathbb{R}$, with $\mathcal{X} \subseteq \mathcal{D}$. Assume that f is differentiable at $\boldsymbol{x}^*$. If $\boldsymbol{x}^*$ is a local maximizer of f in $\mathcal{X}$ then*

$$\nabla f(\boldsymbol{x}^*)^{\mathrm{T}} \boldsymbol{v} \leq 0, \forall \boldsymbol{v} \in \mathcal{T}_{\mathcal{X}}(\boldsymbol{x}^*). \tag{A.31}$$

Proof. [25, Theorem 5.1.2]. □

Definition A.7.5. Let $\boldsymbol{x}^* \in \mathcal{X}$ be a point satisfying (A.31). Then $\boldsymbol{x}^*$ is a *stationary point* in $\mathcal{X}$.

A.8 Differentiable Manifolds

The local method developed in Sect. 3.5 makes use of a number of concepts from differential geometry, namely differentiable manifolds and parameterizations. Within the scope of this monograph, it is sufficient to consider manifolds that are topological subspaces of $\mathbb{R}^M$. Accordingly, in the following, $\mathcal{M}$ always denotes a topological subspace of $\mathbb{R}^M$.

Definition A.8.1 (Local Parameterization). Let $\mathcal{X}$ be a nonempty open subset of $\mathbb{R}^N$ and $\phi : \mathcal{X} \to \mathcal{M}$ a homeomorphism. Then $(\mathcal{X}, \phi)$ is a local parameterization of $\mathcal{M}$.

Definition A.8.2 (Global Parameterization). Let $(\mathcal{X}, \phi)$ be a local parameterization of $\mathcal{M}$. If $\phi(\mathcal{X}) = \mathcal{M}$, then $(\mathcal{X}, \phi)$ is a global parameterization of $\mathcal{M}$.

Definition A.8.3 (Manifold). Let $\{(\mathcal{X}_\alpha, \phi_\alpha)\}_{\alpha\in\mathcal{I}}$ be a collection of local parameterizations of $\mathcal{M}$ such that

$$\mathcal{M} = \bigcup_{\alpha\in\mathcal{I}} \phi_\alpha(\mathcal{X}_\alpha),$$

where $\mathcal{X}_\alpha \subseteq \mathbb{R}^N$. Then $\mathcal{M}$ is a topological manifold of dimension N.

Note that the general definition of a manifold requires that $\mathcal{M}$ be Hausdorff and second-countable [66]. These two properties are automatically fulfilled by any topological subspace of $\mathbb{R}^M$, and are therefore omitted in Definition A.8.3.

Definition A.8.4. A function $\boldsymbol{f} : \mathcal{X} \to \mathbb{R}^M$, with $\mathcal{X} \subseteq \mathbb{R}^N$ open, is of class C^1 if the partial derivatives $\frac{\partial f_m}{\partial x_n} : \mathcal{X} \to \mathbb{R}$ are continuous. The function $\boldsymbol{f}$ is of class C^r, $r \geq 2$, if the partial derivatives $\frac{\partial f_m}{\partial x_n} : \mathcal{X} \to \mathbb{R}$ are of class C^{r-1}.

Definition A.8.5. Two local parameterizations $(\mathcal{X}_\alpha, \phi_\alpha)$ and $(\mathcal{X}_\beta, \phi_\beta)$ are C^r-compatible if $\phi_\alpha(\mathcal{X}_\alpha) \cap \phi_\beta(\mathcal{X}_\beta) \neq \emptyset$ implies that $\phi_\alpha^{-1} \circ \phi_\beta$ is of class C^r.

Definition A.8.6. Let $\mathcal{M}$ be a topological manifold, and let $\mathcal{P} = \{(\mathcal{X}_\alpha, \phi_\alpha)\}_{\alpha\in\mathcal{I}}$ be a collection of local parameterizations that covers $\mathcal{M}$. Moreover, assume that all elements of $\mathcal{P}$ are pairwise C^r compatible. A C^r structure on $\mathcal{M}$ is a collection $\mathcal{D}$ of local parameterizations of $\mathcal{M}$ such that:

1. $\mathcal{D} \supseteq \mathcal{P}$,
2. If $(\mathcal{X}, \phi)$ is a local parameterization of $\mathcal{M}$ that is C^r compatible with all elements of $\mathcal{D}$, then $(\mathcal{X}, \phi)$ is in $\mathcal{D}$.

The C^r structure $\mathcal{D}$ is obtained by adding all C^r compatible parameterizations to $\mathcal{P}$. Accordingly, $\mathcal{D}$ is said to be *generated* by $\mathcal{P}$.

Definition A.8.7 (Differentiable Manifold). Let $\mathcal{M}$ be a topological manifold, and let $\mathcal{D}$ be a C^r structure on $\mathcal{M}$. Then $(\mathcal{M}, \mathcal{D})$ is a differentiable manifold of class C^r.

One of the simplest examples of a differentiable manifold of class C^r is $(\mathbb{R}^M, \mathcal{D})$, where $\mathcal{D}$ is a C^r structure generated by $\mathcal{P} = \{(\mathbb{R}^M, \mathrm{id})\}$.

Lemma A.8.1. *Suppose that $\mathcal{D}$ is a C^r structure on $\mathbb{R}^M$ satisfying $(\mathbb{R}^M, \mathrm{id}) \in \mathcal{D}$. Let $(\mathcal{X}, \phi) \in \mathcal{D}$. Then ϕ is of class C^r.*

Proof. The parameterization $(\mathcal{X}, \phi)$ is C^r compatible with $(\mathbb{R}^M, \mathrm{id})$. By definition, $\mathcal{X}$ is nonempty, hence $\mathcal{X} \cap \mathbb{R}^M \neq \emptyset$, implying that $\mathrm{id}^{-1} \circ \phi = \phi$ is of class C^r. □

In the context of this monograph, parameterizations $(\mathcal{X}, \phi)$ with ϕ of class C^1 are of particular importance, cf. Sect. 3.5. However, the assumption that $(\mathcal{M}, \mathcal{D})$ is a manifold of class C^1 is not sufficient for a parameterization of $\mathcal{M}$ to be of class C^1. As an example, let $(\mathcal{X}, \phi)$ be a global parameterization of $\mathcal{M}$. Then $\{(\mathcal{X}, \phi)\}$ is a collection that covers $\mathcal{M}$ and $\phi^{-1} \circ \phi$ is of class C^∞, hence $(\mathcal{M}, \mathcal{D})$, with $\mathcal{D} \supseteq \{(\mathcal{X}, \phi)\}$, is a differentiable manifold, regardless of the differentiability class of ϕ. A sufficient condition for the existence of a local parameterization such that ϕ is of class C^1 is provided by Proposition A.8.2. Proposition A.8.2 relies on the concept of a regular submanifold, which is developed next.

Definition A.8.8. A subset $\mathcal{N}$ of a C^r manifold $(\mathcal{M}, \mathcal{D})$ has the *N-submanifold property* if for each $\boldsymbol{y} \in \mathcal{N}$ there exists a parameterization $(\mathcal{X}, \boldsymbol{\phi}) \in \mathcal{D}$ such that $\boldsymbol{y} \in \boldsymbol{\phi}(\mathcal{X})$ and

$$\boldsymbol{\phi}^{-1}(\mathcal{N} \cap \boldsymbol{\phi}(\mathcal{X})) \subseteq \mathbb{R}^N \times \{\mathbf{0}\}. \tag{A.32}$$

Parameterizations that fulfill these requirements are called *preferred* parameterizations (relative to $\mathcal{N}$).

If a set $\mathcal{N}$ has the N-submanifold property, it can be parameterized by using the first N coordinates of the preferred parameterizations—according to (A.32), the remaining $M - N$ coordinates are identically to zero, hence do not carry any information. Intuitively, a local parameterization of $\mathcal{N}$ can be obtained by removing the coordinates that are identical to zero. In order to do so, let $M > N$ and define maps $\boldsymbol{\pi}$ and $\boldsymbol{\theta}$ as follows:

$$\boldsymbol{\pi} : \mathbb{R}^M \to \mathbb{R}^N, (x_1, \ldots, x_M) \mapsto (x_1, \ldots, x_N),$$
$$\boldsymbol{\theta} : \mathbb{R}^N \to \mathbb{R}^M, \boldsymbol{y} \mapsto (\boldsymbol{y}, \mathbf{0}).$$

While $\boldsymbol{\pi}$ projects on the first N coordinates of $\boldsymbol{x} \in \mathbb{R}^M$, $\boldsymbol{\theta}$ appends $M - N$ zeros to $\boldsymbol{y} \in \mathbb{R}^N$. Suppose that $\mathcal{N}$ is a topological space that has the N-submanifold property. Let $(\mathcal{X}, \boldsymbol{\phi})$ be a preferred parameterization relative to $\mathcal{N}$. Then $\boldsymbol{\psi} : \mathcal{Y} \to \mathcal{N}$, with

$$\mathcal{Y} = \boldsymbol{\pi}(\boldsymbol{\phi}^{-1}(\mathcal{N} \cap \boldsymbol{\phi}(\mathcal{X}))), \tag{A.33}$$
$$\boldsymbol{\psi}(\boldsymbol{y}) = \boldsymbol{\phi}(\boldsymbol{\theta}(\boldsymbol{y})), \tag{A.34}$$

is a homeomorphism, provided that $\mathcal{N} \cap \boldsymbol{\phi}(\mathcal{X})$ is an open set in the topology of $\mathcal{N}$. The set $\boldsymbol{\phi}(\mathcal{X})$ is open in $\mathcal{M}$, hence $\mathcal{N} \cap \boldsymbol{\phi}(\mathcal{X})$ is open in $\mathcal{N}$ if $\mathcal{N}$ is equipped with the subspace topology. As a result, if $\mathcal{N}$ has the subspace topology then $(\mathcal{Y}, \boldsymbol{\psi})$ is a local parameterization of $\mathcal{N}$. Let $\mathcal{P}$ be the collection of local parameterizations obtained from all the preferred parameterizations. By Definition A.8.8, $\mathcal{P}$ covers $\mathcal{N}$, hence $\mathcal{N}$ equipped with the subspace topology is a topological manifold of dimension N. Moreover, the parameterizations in $\mathcal{P}$ are pairwise C^r compatible. Let $(\mathcal{Y}_\alpha, \boldsymbol{\psi}_\alpha), (\mathcal{Y}_\beta, \boldsymbol{\psi}_\beta) \in \mathcal{P}$ such that $\boldsymbol{\psi}_\alpha(\mathcal{Y}_\alpha) \cap \boldsymbol{\psi}_\beta(\mathcal{Y}_\beta) \neq \emptyset$. Then

$$\boldsymbol{\psi}_\alpha^{-1} \circ \boldsymbol{\psi}_\beta = \boldsymbol{\pi} \circ \boldsymbol{\phi}_\alpha^{-1} \circ \boldsymbol{\phi}_\beta \circ \boldsymbol{\theta},$$

where $(\mathcal{X}_\alpha, \boldsymbol{\phi}_\alpha), (\mathcal{X}_\beta, \boldsymbol{\phi}_\beta) \in \mathcal{D}$ are the corresponding preferred parameterizations. But $\boldsymbol{\phi}_\alpha^{-1} \circ \boldsymbol{\phi}_\beta$ is C^r, and $\boldsymbol{\pi}$ and $\boldsymbol{\theta}$ are C^∞, hence $\boldsymbol{\psi}_\alpha^{-1} \circ \boldsymbol{\psi}_\beta$ is of class C^r. By Definition A.8.6, $\mathcal{P}$ generates a C^r structure on $\mathcal{N}$. In the following, let $\mathcal{F}$ denote the C^r structure generated by $\mathcal{P}$.

Definition A.8.9 (Regular Submanifold). Let $(\mathcal{M}, \mathcal{D})$ be a differentiable manifold of class C^r. Suppose $\mathcal{N} \subset M$ is equipped with the subspace topology and has the N-submanifold property. Then $(\mathcal{N}, \mathcal{F})$ is a *regular submanifold* of $(\mathcal{M}, \mathcal{D})$.

Proposition A.8.2. *Let $(\mathcal{N}, \mathcal{F})$ be a regular submanifold of $(\mathbb{R}^M, \mathcal{D})$, where $\mathcal{D}$ is a C^r structure satisfying $(\mathbb{R}^M, \mathrm{id}) \in \mathcal{D}$. For each $\boldsymbol{s} \in \mathcal{N}$ there exists a local parameterization $(\mathcal{Y}, \psi) \in \mathcal{F}$ such that $\boldsymbol{s} \in \psi(\mathcal{Y})$ and ψ is of class C^r.*

Proof. By Definition A.8.9, for each $\boldsymbol{s} \in \mathcal{N}$ there exists a preferred parameterization $(\mathcal{X}, \phi) \in \mathcal{D}$ such that $\boldsymbol{s} \in \phi(\mathcal{X})$. By (A.33) and (A.34), $(\mathcal{X}, \phi)$ induces a local parameterization $(\mathcal{Y}, \psi)$ of $\mathcal{N}$, with $\psi(\boldsymbol{y}) = \phi(\boldsymbol{\theta}(\boldsymbol{y}))$. By Lemma A.8.1, ϕ is of class C^r, while $\boldsymbol{\theta}$ is of class C^∞, hence the composition of ϕ and $\boldsymbol{\theta}$ is of class C^r. Finally, $\psi(\mathcal{Y}) = \mathcal{N} \cap \phi(\mathcal{X})$, thus $\boldsymbol{s} \in \psi(\mathcal{Y})$. □

By Proposition A.8.2, regular submanifolds of $(\mathbb{R}^M, \mathcal{D})$, where $\mathcal{D}$ is a C^r structure satisfying $(\mathbb{R}^M, \mathrm{id}) \in \mathcal{D}$, have the attractive property that they have parameterizations of class C^r. Due to this property, such manifolds play an important role in the development of a local method in Sect. 3.5.

Definition A.8.10. Let $(\mathbb{R}^M, \mathcal{D})$ be a differentiable manifold of class C^r, with $(\mathbb{R}^M, \mathrm{id}) \in \mathcal{D}$. Suppose $(\mathcal{N}, \mathcal{F})$ is a regular submanifold of $(\mathbb{R}^M, \mathcal{D})$. Then $\mathcal{N}$ is denoted as C^r submanifold of $\mathbb{R}^M$.

Note that when saying that $\mathcal{N}$ is a C^r submanifold of $\mathbb{R}^M$, it is implied that the differentiable structures $\mathcal{D}$ and $\mathcal{F}$ are used, where $\mathcal{D}$ is a C^r structure on $\mathbb{R}^M$ satisfying $(\mathbb{R}^M, \mathrm{id}) \in \mathcal{D}$, and $\mathcal{F}$ is the C^r structure on $\mathcal{N}$ obtained from the preferred parameterizations relative to $\mathcal{N}$.

Proposition A.8.3. *Let $\mathcal{N}$ be an N-dimensional manifold. Furthermore, suppose $\mathcal{N}$ is a C^r submanifold of $\mathbb{R}^M$. The tangent cone of $\mathcal{N}$ at $\boldsymbol{s} \in \mathcal{N}$ is an N-dimensional linear subspace of $\mathbb{R}^M$ and is given by*

$$\mathcal{T}_{\mathcal{N}}(\boldsymbol{s}) = \mathrm{range}\left(\nabla\phi(\boldsymbol{\mu})^{\mathrm{T}}\right),$$

where $(\mathcal{X}, \phi)$ is a local parameterization of $\mathcal{N}$ such that $\boldsymbol{s} = \phi(\boldsymbol{\mu})$ and ϕ is of class C^r.

Proof. Existence of a parameterization $(\mathcal{X}, \phi)$ such that $\boldsymbol{s} \in \phi(\mathcal{X})$ and ϕ is of class C^r follows from Proposition A.8.2. Let $\boldsymbol{\mu} = \phi^{-1}(\boldsymbol{s})$. Due to the fact that ϕ is of class C^r, $\phi(\boldsymbol{\mu}')$ can be written as

$$\phi(\boldsymbol{\mu}') = \phi(\boldsymbol{\mu}) + \nabla\phi(\boldsymbol{\mu})^{\mathrm{T}}(\boldsymbol{\mu}' - \boldsymbol{\mu}) + \boldsymbol{z}(\boldsymbol{\mu}'), \tag{A.35}$$

where $\boldsymbol{z}$ is of class C^r with

$$\nabla \boldsymbol{z}(\boldsymbol{\mu}) = \boldsymbol{0}. \tag{A.36}$$

Let $\{t^{(n)}\}$ be a sequence such that $t^{(n)} \in \mathbb{R}_{++}$ and

$$\lim_{n\to\infty} t^{(n)} = 0.$$

Let

$$\boldsymbol{\mu}^{(n)} = \boldsymbol{\mu} + t^{(n)} \Delta\boldsymbol{\mu},$$

with $\Delta\boldsymbol{\mu} \in \mathbb{R}^N$. For n large enough $\boldsymbol{\mu}^{(n)} \in \mathcal{X}$. Moreover, by continuity of ϕ,

$$\lim_{n\to\infty} \phi(\boldsymbol{\mu}^{(n)}) = \phi(\boldsymbol{\mu}).$$

Plugging (A.35) into (A.30) yields

$$\nabla\phi(\boldsymbol{\mu})^{\mathrm{T}}\Delta\boldsymbol{\mu} + \lim_{n\to\infty} \frac{\boldsymbol{z}(\boldsymbol{\mu}^{(n)})}{t^{(n)}} = \boldsymbol{v}. \tag{A.37}$$

By (A.36) and l'Hôpital's rule,

$$\lim_{n\to\infty} \frac{\boldsymbol{z}(\boldsymbol{\mu}^{(n)})}{t^{(n)}} = \mathbf{0}.$$

Moreover, (A.37) holds for any $\Delta\boldsymbol{\mu} \in \mathbb{R}^N$, hence

$$\mathcal{T}_{\mathcal{N}}(\boldsymbol{s}) = \text{range}\left(\nabla\phi(\boldsymbol{\mu})^{\mathrm{T}}\right).$$

The map ϕ is a homeomorphism, hence rank $\left(\nabla\phi(\boldsymbol{\mu})^{\mathrm{T}}\right) = N$. □

Proposition A.8.4. *Let $\mathcal{N}$ be an $M-1$-dimensional C^r submanifold of $\mathbb{R}^M$. There exists a continuous map $\boldsymbol{n} : \mathcal{N} \to \mathbb{R}^M$ such that*

$$\|\boldsymbol{n}(\boldsymbol{s})\|_2 = 1,$$
$$\mathcal{T}_{\mathcal{N}}(\boldsymbol{s}) = \text{null}\left(\boldsymbol{n}(\boldsymbol{s})^{\mathrm{T}}\right).$$

Proof. By Proposition A.8.3, $\mathcal{T}_{\mathcal{N}}(\boldsymbol{s})$ is an $M-1$ dimensional subspace of $\mathbb{R}^M$. Hence, at each $\boldsymbol{s} \in \mathcal{N}$, there exists $\boldsymbol{n} \in \mathbb{R}^M$ such that

$$\mathcal{T}_{\mathcal{N}}(\boldsymbol{s}) = \text{null}\left(\boldsymbol{n}^{\mathrm{T}}\right).$$

Moreover, at each $\boldsymbol{s} \in \mathcal{N}$ there exists a local parameterization $(\mathcal{X}, \phi)$ such that

$$\text{range}\left(\nabla\phi(\phi^{-1}(\boldsymbol{s}))^{\mathrm{T}}\right) = \text{null}\left(\boldsymbol{n}^{\mathrm{T}}\right).$$

Equivalently, $\boldsymbol{n}$ is a nontrivial solution of the following system of equations:

$$\nabla\phi(\phi^{-1}(\boldsymbol{s}))\boldsymbol{n} = \mathbf{0}. \tag{A.38}$$

The parameterization ϕ in Proposition A.8.3 is of class C^r, thus $\nabla\phi \circ \phi^{-1}$ is continuous. As a result, it is possible to find a continuous map $\boldsymbol{n} : \mathcal{N} \to \mathbb{R}^M$ such that $\boldsymbol{n}(\boldsymbol{s})$ is a nontrivial solution of (A.38). Moreover, if $\boldsymbol{n}$ is a solution of (A.38), then $\boldsymbol{n}/\|\boldsymbol{n}\|_2$ is also a solution. The map $\omega : \boldsymbol{x} \mapsto \boldsymbol{x}/\|\boldsymbol{x}\|_2$ is continuous on $\mathbb{R}^M \setminus \{\mathbf{0}\}$, hence $\omega \circ \boldsymbol{n}$ is continuous. □

A.9 Yates' Framework for Power Control

Results from Yates' work [2] are repeatedly used throughout this monograph. For completeness, this appendix reproduces the relevant definitions and results from [2].

Definition A.9.1 (Standard Interference Function). A function $\boldsymbol{f} : \mathbb{R}_+^K \to \mathbb{R}_+^K$ is a *standard interference function* if it satisfies the following properties:

$$\begin{aligned} \boldsymbol{f}(\boldsymbol{p}) &\gg \boldsymbol{0}, \\ \boldsymbol{p} \geq \boldsymbol{p}' &\implies \boldsymbol{f}(\boldsymbol{p}) \geq \boldsymbol{f}(\boldsymbol{p}'), \\ \alpha \boldsymbol{f}(\boldsymbol{p}) &\gg \boldsymbol{f}(\alpha \boldsymbol{p}'), \forall \alpha > 1. \end{aligned}$$

Theorem A.9.1. *Let $\boldsymbol{f}$ be a standard interference function. If there exists $\boldsymbol{p}^* \in \mathbb{R}_+^K$ such that*

$$\boldsymbol{p}^* = \boldsymbol{f}(\boldsymbol{p}^*),$$

then $\boldsymbol{p}^*$ is unique.

Lemma A.9.2. *Let $\boldsymbol{f}$ be a standard interference function and define a sequence $\{\boldsymbol{p}^{(n)}\}$ as follows:*

$$\boldsymbol{p}^{(n+1)} = \boldsymbol{f}(\boldsymbol{p}^{(n)}).$$

Assume that $\boldsymbol{p}^{(0)} \geq \boldsymbol{f}(\boldsymbol{p}^{(0)})$. Then

$$\begin{aligned} \boldsymbol{p}^{(n+1)} &\leq \boldsymbol{p}^{(n)}, \\ \boldsymbol{p}^{(n)} &\geq \boldsymbol{f}(\boldsymbol{p}^{(n)}), \\ \lim_{n\to\infty} \boldsymbol{p}^{(n)} &= \boldsymbol{p}^*, \end{aligned}$$

where $\boldsymbol{p}^*$ is a unique fixed point such that $\boldsymbol{p}^* = \boldsymbol{f}(\boldsymbol{p}^*)$.

Corollary A.9.3. *Assume there exists $\boldsymbol{p}^*$ such that $\boldsymbol{p}^* = \boldsymbol{f}(\boldsymbol{p}^*)$. Then*

$$\boldsymbol{p} > \boldsymbol{f}(\boldsymbol{p}) \implies \boldsymbol{p} > \boldsymbol{p}^*.$$

References

1. Rappaport, T.: Wireless Communications: Principles and Practice. 2nd edn. Prentice Hall, New Jersey (2001)
2. Yates, R.: A framework for uplink power control in cellular radio systems. IEEE J. Sel. Area. Comm. **13**(7), 1341–1347 (1995)
3. Stanczak, S., Wiczanowski, M., Boche, H.: Fundamentals of resource allocation in wireless networks: Theory and algorithms. In: Utschick, W., Boche, H., Mathar, R. (eds.) ser. Foundations in Signal Processing, Communications and Networking, 2nd edn. Springer, New York (2009)
4. Goldmedia GmbH: BITKOM Goldmedia mobile life report 2012, (2008). http://www.bitkom.org/files/documents/081009_BITKOM_Goldmedia_Mobile_Life_2012.pdf
5. Kelly, F.: Charging and rate control for elastic traffic. Eur. Trans. Telecomm. **8**, 33–37 (1997)
6. Shenker, S.: Fundamental design issues for the future internet. IEEE J. Sel. Area. Comm. **13**(7), 1176–1188 (1995)
7. Cao, Z., Zegura, E.: Utility max–min: An application-oriented bandwidth allocation scheme. In: Proceedings of the IEEE Conference on Computer Communications (INFOCOM), vol. 2, pp. 793–801 (1999)
8. Schwarz, H., Marpe, D., Wiegand, T.: Overview of the scalable video coding extension of the H.264/AVC standard. IEEE Trans. Circ. Syst. Video Tech. **17**, 1136–1148 (2007)
9. van Der Schaar, M., Sai Shankar, N: Cross-layer wireless multimedia transmission: Challenges, principles, and new paradigms. IEEE Wireless Comm. **12**, 50–58 (2005)
10. Srivastava, V., Motani, M.: Cross-layer design: A survey and the road ahead. IEEE Comm. Mag. **43**(12), 112–119 (2005)
11. Katsaggelos, A.K., Eisenberg, Y., Zhai, F., Berry, R., Pappas, T.N.: Advances in efficient resource allocation for packet-based real-time video transmission. Proc. IEEE **93**, 135–147 (2005)
12. Khan, S., Peng, Y., Steinbach, E., Sgroi, M., Kellerer, W.: Application-driven cross-layer optimization for video streaming over wireless networks. IEEE Comm. Mag. **44**, 122–130 (2006)
13. Lin, X., Shroff, N.B., Srikant, R.: A tutorial on cross-layer optimization in wireless networks. IEEE J. Sel. Area. Comm. **24**, 1452–1463 (2006)
14. Chiang, M., Low, S.H., Calderbank, A.R., Doyle, J.C.: Layering as optimization decomposition: A mathematical theory of network architectures. Proc. IEEE **95**, 255–312 (2007)
15. Akyildiz, I.F., Wang, X.: Cross-layer design in wireless mesh networks. IEEE Trans. Veh. Technol. **57**(2), 1061–1076 (2008)
16. Foukalas, F., Gazis, V., Alonistioti, N.: Cross-layer design proposals for wireless mobile networks: A survey and taxonomy. IEEE Comm. Surv. Tutorials **10**(1), 70–85 (2008)

J. Brehmer, *Utility Maximization in Nonconvex Wireless Systems*, Foundations in Signal Processing, Communications and Networking 5, DOI 10.1007/978-3-642-17438-4,

17. Chiang, M.: Balancing transport and physical layers in wireless multihop networks: jointly optimal congestion control and power control. IEEE J. Sel. Area. Comm. **23**, 104–116 (2005)
18. Stanczak, S., Wiczanowski, M., Boche, H.: Distributed utility-based power control: Objectives and algorithms. IEEE Trans. Signal Process. **55**(10), 5058–5068 (2007)
19. Chiang, M.: Nonconvex optimization in communication systems. In: Gao, D.Y., Sherali, H.D. (eds.) In: Advances in Applied Mathematics and Global Optimization: In: Honor of Gilbert Strang, ser. Advances in Mechanics and Mathematics. vol. 17., Springer, New York (2009)
20. Vavasis, S.A.: Complexity issues in global optimization: A survey. In: Handbook of Global Optimization. pp. 27–41. Kluwer, Dordrecht (1995)
21. Zimmermann, H.: OSI reference model – the ISO model of architecture for open systems interconnection. IEEE Trans. Comm. **28**(4), 425–432 (1980)
22. Hiriart-Urruty, J., Lemaréchal, C.: Convex analysis and minimization algorithms I. ser. Grundlehren der Mathematischen Wissenschaften. Vol. 305, Springer, Berlin (1996)
23. Ben-Tal, A., Nemirovski, A.: Lectures on modern convex optimization: Analysis, algorithms, and engineering applications. Society for Industrial and Applied Mathematics, Philadelphia, PA, USA (2001)
24. Tuy, H.: Monotonic optimization: Problems and solution approaches. SIAM J. Optim. **11**(2), 464–494 (2000)
25. Bazaraa, M.S., Sherali, H.D., Shetty, C.M.: Nonlinear Programming. Wiley, Hoboken, NJ (2006)
26. Palomar, D.P., Chiang, M.: A tutorial on decomposition methods for network utility maximization. IEEE J. Sel. Area. Comm. **24**, 1439–1451 (2006)
27. Palomar, D.P., Chiang, M.: Alternative distributed algorithms for network utility maximization: Framework and applications. IEEE Trans. Automat. Contr. **52**(12), 2254–2269 (2007)
28. Bertsekas, D., Nedic, A., Ozdaglar, A.: Convex Analysis and Optimization. Athena Scientific, Belmont, MA (2003)
29. Hiriart-Urruty, J., Lemaréchal, C.: Convex Analysis and Minimization Algorithms II, ser. Grundlehren der mathematischen Wissenschaften. vol. 306, Springer, Berlin (1996)
30. Goffin, J.-L., Vial, J.-P.: Convex nondifferentiable optimization: A survey focused on the analytic center cutting plane method. Tech. Rep. 99.02, Logilab, Geneva, Switzerland (1999)
31. Kelly, F., Maulloo A.K., Tan, D.: Rate control for communication networks: Shadow prices, proportional fairness and stability. J. Oper. Res. Soc. **49**(3), 237–252 (1998)
32. Liu, J., Hou, Y.T., Sherali, H.D.: Routing and power allocation for MIMO-based ad hoc networks with dirty paper coding. In: Proceedings of the IEEE International Conference on Communications (ICC). pp. 2859–2864 (2008)
33. Liu, J., Hou, Y.T.: Weighted proportional fairness capacity of Gaussian MIMO broadcast channels. In: Proceedings of the IEEE Conference on Computer Communications (INFOCOM), pp. 385–393 (2008)
34. Lee, J., Jindal, N.: Symmetric capacity of MIMO downlink channels. In: Proceedings of the IEEE International Symposium on Information Theory (ISIT), pp. 1031–1035 (2006)
35. Michel, T., Wunder, G.: Achieving QoS and efficiency in the MIMO downlink with limited power. In: Proceedings of the ITG/IEEE International Workshop on Smart Antennas (WSA). Vienna, Austria (2007)
36. Lee, J.W., Mazumdar, R.R., Shroff, N.B.: Non-convex optimization and rate control for multi-class services in the internet. IEEE/ACM Trans. Netw. **13**(4), 827–840 (2005)
37. Nedic, A., Ozdaglar, A.: Approximate primal solutions and rate analysis for dual subgradient methods. Tech. Report LIDS 2753, Massachusetts Institute of Technology (2007)
38. Hande, P., Shengyu, Z., Chiang, M.: Distributed rate allocation for inelastic flows. IEEE/ACM Trans. Netw. **15**(6), 1240–1253 (2007)
39. Lee, J.W., Mazumdar, R.R., Shroff, N.B.: Downlink power allocation for multi-class wireless systems. IEEE/ACM Trans. Netw. **13**, 854–867 (2005)
40. Tuy, H., Al-Khayyal, F., Thach, P.: Monotonic optimization: Branch and cut methods. In: Audet, C., Hansen, P., Savard, G. (eds.) Essays and Surveys in Global Optimization. pp. 39–78. Springer, USA (2005)

41. Jorswieck, E., Larsson, E.: Linear precoding in multiple antenna broadcast channels: Efficient computation of the achievable rate region. In: Proceedings of the ITG/IEEE International Workshop on Smart Antennas (WSA). Aachen, Germany (2008)
42. Brehmer, J., Utschick, W.: Utility maximization in the multi-user MISO downlink with linear precoding. In: Proceedings of the IEEE International Conference on Communications (ICC). Dresden, Germany (2009)
43. Qian, L.P., Zhang, Y.J., Huang, J.: MAPEL: Achieving global optimality for a non-convex wireless power control problem. IEEE Trans. Wireless Commun. **8**(3), 1553–1563 (2009)
44. Brehmer, J., Utschick, W.: Nonconcave utility maximisation in the MIMO broadcast channel. EURASIP J. Adv. Signal Process. **2009**(1), 645041 (2009). http://asp.eurasipjournals.com/content/2009/1/645041, doi: 10.1155/2009/645041
45. Jorswieck, E., Larsson, E.: Monotonic optimization framework for the MISO IFC. In: Proceedings of the IEEE International Conference on Acoustics, Speech, and Signal Processing (ICASSP). pp. 3633–3636 (2009)
46. Brehmer, J., Utschick, W.: On proportional fairness in nonconvex wireless systems. In: Proceedings of the ITG/IEEE International Workshop on Smart Antennas (WSA), Berlin, Germany (2009)
47. Jorswieck, E., Larsson, E.: The MISO interference channel from a game-theoretic perspective: A combination of selfishness and altruism achieves Pareto optimality. In: Proceedings of the IEEE International Conference on Acoustics, Speech, and Signal Processing (ICASSP) pp. 5364–5367 (2008)
48. Nokleby, M., Swindlehurst, A.: Bargaining and the MISO interference channel. EURASIP J. Adv. Signal Process. **2009**(1), 368547 (2009). http://asp.eurasipjournals.com/content/2009/1/368547, doi: 10.1155/2009/368547
49. Bengtsson, M., Ottersten, B.: Optimal downlink beamforming using semidefinite optimization. Royal Institute of Technology (KTH). Tech. Report IR-S3-SB-9955, Stockholm, Sweden (1999)
50. Brehmer, J., Utschick, W.: Optimal interference management in multi-antenna, multi-cell systems. In: Proceedings of the International Zurich Seminar on Communications (IZS). Zürich, Switzerland (2010)
51. Fazel, M., Chiang, M.: Network utility maximization with nonconcave utilities using sum-of-squares method. In: Proceedings of the IEEE Conference on Decision and Control (CDC). pp. 1867–1874 (2005)
52. Parrilo, P.: Semidefinite programming relaxations for semialgebraic problems. Math. Program. Series B **96**(2), 293–320 (2003)
53. Boche, H., Wiczanowski, M., Stanczak, S.: Characterization of optimal resource allocation in cellular networks. In: Proceedings of the IEEE International Workshop on Signal Processing Advances for Wireless Communications (SPAWC) pp. 454–458 (2004)
54. Chiang, M., Tan, C.W., Palomar, D.P., O'Neill, D., Julian, D.: Power control by geometric programming. IEEE Trans. Wireless Comm. **6**(7), 2640–2651 (2007)
55. Tan, C.W., Palomar, D.P., Chiang, M.: Energy-robustness tradeoff in cellular network power control. IEEE/ACM Trans. Netw. **17**(3), 912–925 (2009)
56. Mo, J., Walrand, J.: Fair end-to-end window-based congestion control. IEEE/ACM Trans. Netw. **8**(5), 556–567 (2000)
57. Wang, W.-H., Palaniswami, M., Low, S.H.: Application-oriented flow control: Fundamentals, algorithms and fairness. IEEE/ACM Trans. Netw. **14**(6), 1282–1291 (2006)
58. Harks, T., Poschwatta, T.: Congestion control in utility fair networks. Comput. Netw. **52**(15), 2947–2960 (2008)
59. Gabay, D.: Minimizing a differentiable function over a differentiable manifold. J. Optim. Theor. Appl. **37**, 177–219 (1982)
60. Luenberger, D.G.: Linear and Nonlinear Programming, 2nd edn. Addison-Wesley, Inc., Reading, Massachusetts (1984)
61. Rosen, J.B.: The gradient projection method for nonlinear programming, II: Nonlinear constraints. SIAM J. Appl. Math. **9**, 514–553 (1961)

62. Polak, E.: On the convergence of optimization algorithms. Revue franaise d'informatique et de recherche oprationnelle **3**(1), 17–34 (1969)
63. Powell, M.J.D.: Convergence properties of algorithms for nonlinear optimization. SIAM Rev. **28**(4), 487–500 (1986)
64. Du, D.-Z., Zhang, X.-S.: Global convergence of Rosen's gradient projection method. Math. Program. **44**(1), 357–366 (1989)
65. Du, D.-Z., Wu, F., Zhang, X.-S.: On Rosen's gradient projection methods. Ann. Oper. Res. **24**(1), 9–28 (1990)
66. Boothby, W.: An Introduction to Differentiable Manifolds and Riemannian Geometry. Academic Press, New York (1975)
67. Absil, P.-A., Mahony, R., Sepulchre, R.: Optimization Algorithms on Matrix Manifolds. Princeton University Press, Princeton, NJ (2008)
68. Luenberger, D.G.: The gradient projection method along geodesics. Manag. Sci. **18**, 620–631 (1972)
69. Manton, J.H.: On the various generalisations of optimisation algorithms to manifolds. In: Proceedings of the International Symposium on Mathematical Theory of Networks and Systems (MTNS). Leuven, Belgium (2004)
70. Hüper, K., Trumpf, J.: Newton-like methods for numerical optimization on manifolds. In: Proceedings of the Asilomar Conference on Signals, Systems and Computers. vol. 1, pp. 136–139 (2004)
71. Riemensberger, M., Utschick, W.: Gradient descent on differentiable manifolds. Technische Universität München, Tech. Report MSV-2008-1, Munich, Germany (2008)
72. Costa, M.: Writing on dirty paper. IEEE Trans. Inform. Theor. **29**(3), 439–441 (1983)
73. Schubert, M., Boche, H.: Solution of the multiuser downlink beamforming problem with individual SINR constraints. IEEE Trans. Veh. Tech. **53**(1), 18–28 (2004)
74. Weingarten, H., Steinberg, Y., Shamai, S.S.: The capacity region of the Gaussian multiple-input multiple-output broadcast channel. IEEE Trans. Inform. Theor. **52**, 3936–3964 (2006)
75. Tse, D., Hanly, S.: Multiaccess fading channels. I. polymatroid structure, optimal resource allocation and throughput capacities. IEEE Trans. Inform. Theor. **44**(7), 2796–2815 (1998)
76. Vishwanath, S., Jindal, N., Goldsmith, A.: Duality, achievable rates, and sum-rate capacity of Gaussian MIMO broadcast channels. IEEE Trans. Inform. Theor. **49**, 2658–2668 (2003)
77. Viswanathan, H., Venkatesan, S., Huang, H.: Downlink capacity evaluation of cellular networks with known-interference cancellation. IEEE J. Sel. Area. Comm. **21**, 802–811 (2003)
78. Kobayashi, M., Caire, G.: An iterative water-filling algorithm for maximum weighted sum-rate of Gaussian MIMO-BC. IEEE J. Sel. Area. Comm. **24**, 1640–1646 (2006)
79. Böhnke, R., Kammeyer, K.: Weighted sum rate maximization for the MIMO-downlink using a projected conjugate gradient algorithm. In: International Workshop on Cross Layer Design (IWCLD). Jinan, China (2007)
80. Hunger, R., Schmidt, D.A., Joham, M., Utschick, W.: A general covariance-based optimization framework using orthogonal projections. In: Proceedings of the International Workshop on Signal Processing Advances in Wireless Communications (SPAWC). Recife, Brazil (2008)
81. Massoulie, L., Roberts, J.: Bandwidth sharing: Objectives and algorithms. IEEE/ACM Trans. Netw. **10**(3), 320–328 (2002)
82. Mingbo, X., Shroff, N., Chong, E.: A utility-based power-control scheme in wireless cellular systems. IEEE/ACM Trans. Netw. **11**(2), 210–221 (2003)
83. Touati, C., Kameda, H., Inoie, A.: Fairness in non-convex systems. Tech. Report CS-TR-05-4, University of Tsukuba, Ibaraki, Japan (2005)
84. Knopp, R., Humblet, P.: Information capacity and power control in single-cell multiuser communications. In: Proceedings of the IEEE International Conference on Communications (ICC), vol. 1, pp. 331–335 (1995)
85. Viro, O.Y., Ivanov, O.A., Netsvetaev, N.Y., Kharlamov, V.: Elementary topology: Problem textbook. American Mathematical Society, Providence, RI (2008)

Index

J. Brehmer, *Utility Maximization in Nonconvex Wireless Systems*, Foundations in Signal Processing, Communications and Networking 5, DOI 10.1007/978-3-642-17438-4,

MIX
Papier aus verantwortungsvollen Quellen
Paper from responsible sources
FSC® C105338

FSC
www.fsc.org

If you have any concerns about our products,
you can contact us on
ProductSafety@springernature.com

In case Publisher is established outside the EU,
the EU authorized representative is:
Springer Nature Customer Service Center GmbH
Europaplatz 3, 69115 Heidelberg, Germany

Printed by Libri Plureos GmbH
in Hamburg, Germany